Evolution of Animal Behavior

EVOLUTION
OF
ANIMAL BEHAVIOR

Paleontological and Field Approaches

Edited by
MATTHEW H. NITECKI
Field Museum of Natural History

and
JENNIFER A. KITCHELL
University of Michigan

New York   Oxford
OXFORD UNIVERSITY PRESS
1986

Oxford University Press

Oxford   New York   Toronto
Delhi   Bombay   Calcutta   Madras   Karachi
Petaling Jaya   Singapore   Hong Kong   Tokyo
Nairobi   Dar es Salaam   Cape Town
Melbourne   Auckland

and associated companies in
Beirut   Berlin   Ibadan   Nicosia

Published by Oxford University Press, Inc.,
200 Madison Avenue, New York, New York 10016

Oxford is a registered trademark of Oxford University Press

Library of Congress Cataloging-in-Publication Data
Main entry under title:
Evolution of animal behavior.
Papers presented at the 7th Spring Systematics Symposium
of the Field Museum of Natural History, Chicago, May 1984.
Bibliography: p.   Includes index.
1. Animal behavior—Congresses.   2. Behavior evolution—Congresses.
I. Nitecki, Matthew H. II.   Kitchell, Jennifer A.   III. Spring Systematics Symposium
(7th : 1984 : Chicago, Ill.)   IV. Field Museum of Natural History.
QL750.E95   1986   591.5′1   85-28495
ISBN 0-19-504006-6

2 4 6 8 10 9 7 5 3 1

Printed in the United States of America

Preface

This book represents the results of the Field Museum's Seventh Spring Systematics Symposium on the Evolution of Behavior. The symposium was held in May 1984 and was supported in part by the National Science Foundation.

We are grateful to the following who offered counsel and criticism and reviewed manuscripts: Stuart A. Altmann, James S. Ashe, Herbert R. Barghusen, John R. Bolt, Jack Fooden, Harry W. Greene, Antoni Hoffman, James A. Hopson, Michael LaBarbera, David J. Ligon, Gordon H. Orians, Bruce D. Patterson, Colin Patterson, Frank A. Pitelka, Donald S. Sade, J. John Sepkoski, Jr., Robert Trivers, Richard Wragham, and members of the Spring Systematics Committee. Additional thanks are due to the Field Museum and the National Science Foundation Grant DEB-83-01236 for financial support, Zbigniew Jastrzebski for help with illustrative material, and Martha S. Bryant, who was responsible for many editorial chores and for the preparation of the entire volume.

Chicago
September 1985

M. H. N.
J. A. K.

Contents

Contributors

Jeanne Altmann. Allee Laboratory of Animal Behavior, Department of Biology, University of Chicago, 940 East 57th Street, Chicago, Illinois 60637.

John W. Fitzpatrick. Department of Zoology, Field Museum of Natural History, Roosevelt Road at Lake Shore Drive, Chicago, Illinois 60605.

Jennifer A. Kitchell. Museum of Paleontology, University of Michigan, Ann Arbor, Michigan 48109.

George V. Lauder, Jr. Department of Developmental and Cell Biology, University of California, Irvine, California 92717.

Matthew H. Nitecki. Department of Geology, Field Museum of Natural History, Roosevelt Road at Lake Shore Drive, Chicago, Illinois 60605.

John H. Ostrom. Peabody Museum of Natural History, Yale University, P.O. Box 6666, 170 Whitney Avenue, New Haven, Connecticut 06511.

Adolf Seilacher. Institut und Museum für Geologie und Paläontologie, Universität Tübingen, Sigwartstrasse 10, 7400 Tübingen 1, FDR.

Randy Thornhill. Department of Biology, University of New Mexico, Albuquerque, New Mexico 87131.

Glen E. Woolfenden. Department of Biology, University of South Florida, Tampa, Florida 33620.

Evolution of Animal Behavior

Introduction: Evolution and Behavior

Jennifer A. Kitchell and Matthew H. Nitecki

*I*N 1958, when *Behavior and Evolution* (Roe and Simpson, 1958) was published, it is fair to say that most behavioral research was not being carried out within an evolutionary framework. The dual goal of contributors, notably Simpson and Mayr, was to reintroduce evolutionary explanation into behavioral studies and reincorporate behavior into a then genetically dominated view of evolution. The present volume, quite different in content, reflects the subsequent shift of behavioral research toward explicit inclusion of evolutionary theory, yet resembles the earlier volume in the wilful bringing together of contributors from both paleobiology and neontology.

This volume highlights current advances in the general field of behavioral research by focusing on a selected series of studies that confront wide-ranging issues: sexual selection, mate choice, differential parental investment, apparent altruism and cooperative behavior, apparent nonaltruism and parent-offspring conflict, and the relevance of phylogenetic constraints and historical information. An underlying theme is to view both individual and social behaviors in terms of alternate strategies: resource harvesting strategies (Seilacher), life history and energy allocation strategies (Kitchell), parent-offspring-sibling strategies (Fitzpatrick and Woolfenden, Ostrom), and mate selection and parental investment strategies (Thornhill, Altmann). A recurrent query is, What proximate mechanisms initiate conflicts of interest and what evolutionary consequences ensue from maintained conflicts of interest?

The papers in this volume are aimed at the nonspecialist. The selection of topics is inevitably limited and much valuable work is regrettably omitted. The restricted selection reflects the constraints of a one-day symposium format, advisedly designed for lengthy treatments of selected subjects in lieu of a broader but more tightly compressed menu of short presentations.

Contributions fall into two categories: (1) historical approaches to the evolution of behavior in which attention is directed to the antecedent condition (Lauder, Ostrom, Seilacher, Kitchell), and (2) nonhistorical approaches in which attention is directed more toward selection pressures (Thornhill, Fitzpatrick and Woolfenden, Altmann, also Kitchell). The historical approach permits an assessment of patterns of behavioral evolution over time. The nonhistorical approach emphasizes the sufficiency of evolutionary explanation to account for present-day patterns.

The issue of phylogenetic constraint is thoughtfully analyzed by Lauder, whose interest is directed toward the acquisition of evolutionary novelty. Are critical novelties first functional or first behavioral? Lauder emphasizes that any rigorous testing of historical hypotheses, including those on the evolution of behavior, first require corroborated hypotheses of phylogeny. This emphasis differs from those of Simpson (1958) and Mayr (1958) who argued for "use of behavioral characters in taxonomy." Lauder's interest is less in the usefulness of behavioral traits for systematics than in the usefulness of corroborated phylogenetic hypotheses for the working out of behavioral evolution. The case example is the evolution of novel feeding behavior in the centrarchid fishes. The central question raised by Lauder's comparative, cladistic approach is, Can homologous behaviors have a nonhomologous substrate? Specifically, what is the degree of historical congruency among behaviors, functions, neural pathways, and structural morphology?

Large-scale or macroevolutionary patterns of behavior, derived from exceptional fossil records of behavior, provide temporal evidence of behavioral evolution, enabling one to ask whether fixed kinds of behavior persist over evolutionarily long periods of time, and at what rates and in what directions these behavioral programs evolve. The foraging and burrowing behavioral programs of marine sediment-ingesting organisms provide such an exceptional record, although homology cannot be assessed. Seilacher presents evidence that the behavioral program itself is an ancient one, particularly in stable marine environments, with a record extending from the Cambrian through the Tertiary. Such a temporal record supplies information less on the "learning" of the program than on its execution.

Ostrom reviews the recent evidence in support of social behavior among dinosaurs. Cranial display structures and evidence of vocalization capacities in ornithischian reptiles, for example, have suggested that certain groups may have evolved a social behavior. Fossil evidence of juveniles and nests suggests parent-offspring vocalization and parental care. All such evidence is necessarily indirect.

Kitchell examines selective predation and alternate life-history and energy allocation strategies of prey in the naticid gastropod-bivalve interaction. This system is unusual in that direct evidence is retrievable from the fossil record. Kitchell consequently provides historical evidence of behavioral stereotypy over evolutionary time spans. Inefficient components of predatory behavior also persist over similar time spans.

Thornhill's contribution addresses the controversial topic of sexual selection. In what ways do female choice, sexual selection, and differential parental investment in offspring result in various mating systems in insects? Coupling field observations with experimental work, particularly on *Panorpa* (scorpionflies), Thornhill provides data on the differential consequences on fitness associated with mate choice. A major topic is the controversial one of the role played by female choice in the evolution of male traits. Thornhill examines the interactive role of developmental, genetic, and social constraints on the maintenance of alternative mating strategies. As in Kitchell's study of a biotic environment of selective predation, selective mating criteria can result in non-random differential reproduction, among both males and females. As in the case of the Florida Scrub Jay, resource scarcity is an important ecological determinant of strategy.

Fitzpatrick and Woolfenden study a reproductive system that outwardly manifests cooperation, namely cooperative breeding in the Florida Scrub Jay. These authors more clearly delineate causal interpretations of life history strategy framed in terms of kin selection theory and causality based on individual selection. They conclude that although kin selection is an important component in the evolution of helping behavior, it is not the only component, and, moreover, that there are direct, albeit delayed, benefits to helpers measured in terms of inheriting a breeding space territory. Short-term *altruistic* behavior is maintained because it enhances the probability of long-term individual reproductive success. Siblings, acting as nest helpers, are not fraternal altruists; the behavior of helping, viewed over a longer time frame, increases the probability of individual reproduction, making it an example of "selfishness through cooperation."

Ecological variables and behavioral strategies are again seen to be interdependent. If cooperative breeding behavior is ecologically constrained by a saturated habitat offering limited access to breeding territory, then the defense of suitable nesting territory may be causally related to the evolution of social behaviors, from monogamous pairing to cooperative breeding to plural breeding. As with Altmann, longitudinal studies provide the requisite information on genetic relationships between individuals and long-term reproductive success.

Parental behavior presents no real evolutionary paradox, yet there is the potential for conflict between parent and current vs. future offspring. Excessive parental care of any one offspring may incur a parental reproductive cost at a future date. Altmann evaluates these theoretical ideas with empirical evidence from long-term studies of natural populations of baboons. As with cooperative breeding, curtailed care of current offspring provides a delayed or future benefit in terms of long-term reproductive success. There is also a scope of parental strategies, each with fitness consequences.

Altmann's assessment of the degree of resolution between evolutionary theory and parent-offspring interactions emphasizes the mother-offspring relationship, although there is some information on adult male-offspring (paternity is uncertain) interactions. Because males disperse, the potential for reciprocal altruism involving males is diminished.

As this volume demonstrates, the evolution of behavior is a topic of mutual interest to paleobiologists, systematists, functional morphologists, ecologists, and ethologists, dealing as they do with different time scales, taxonomic levels, and means of approach to testing, including observation, field and laboratory experiments, and pattern analysis. As this volume also connotes, there currently is a healthy mix of approaches to the same problem. In our opinion, the volume will succeed if it arouses in the reader the desire to find out more.

LITERATURE CITED

Mayr, E. 1958. Behavior and systematics. In: Roe, A., and G. G. Simpson (eds), Behavior and Evolution. Yale University Press, New Haven, Conn., pp. 341–362.

Roe, A., and G. G. Simpson (eds). 1958. Behavior and Evolution. Yale University Press, New Haven, Conn., 557 pp.

Simpson , G. G. 1958. Behavior and evolution: Methods and present status of theory. In: Roe, A., and G. G. Simpson (eds). Behavior and Evolution. Yale University Press, New Haven, Conn., pp. 507–535.

I

HISTORICAL APPROACHES TO THE EVOLUTION OF BEHAVIOR

1

Homology, Analogy, and the Evolution of Behavior

George V. Lauder

A NY comparative study in biology of necessity involves some attempt to assess similarities and differences among organisms. These similarities and differences may be interpreted within two general frameworks: (1) an *equilibrium* view in which behaviors or morphologies are viewed in relation to the environment and explanations of similarities and differences are sought in terms of selection forces and environmental variables, and (2) a *transformational* view where patterns of similarity and difference among organisms, species, and higher taxa may be used to construct evolutionary sequences. The transformational approach (Lauder, 1981) focuses on historical patterns of structure, function, and behavior, and relies fundamentally on phylogenetic hypotheses to provide a basis for interpreting historical sequences and testing evolutionary concepts. The subject of this volume is the evolution of behavior, and in this chapter I will focus on historical interpretations of similarities and differences in behavior, and on methods for analyzing evolutionary patterns of behavior.

There has been an interest in transformational patterns of behavior and in the comparison of behaviors in relation to the phylogenetic relationships of species from the early days of ethology (Heinroth, 1911; Whitman, 1899). Lorenz (1950, p. 238), for example, remarked that "as early as 1898, C. O. Whitman wrote the sentence that marks the birth of comparative ethology. 'Instincts and organs are to be studied from the common viewpoint of phyletic descent.'" However, most ethologists in the last twenty years have focused on the relationship of behavior to the environment inhabited by organisms, and this focus is well exemplified by the papers in this book (e.g., Fitzpatrick and Woolfenden, Thornhill). The dominant emphasis of ethologists has been on the physiological basis of behavior (see Fentress, 1976; Manning, 1967) and on the equilibrium relationship of behaviors to fitness, selection pressures, and environments. For example, Manning (1967, p. 1) opens his text on animal behavior with the statement that "there are two main approaches to the study of behavior, the physiological and the psychological." A survey of a number of texts in ethology reveals that only two (Alcock, 1975; Eibl-Eibesfeldt, 1970) even list the words "phylogeny" and/or "homology" in the index (e.g., Manning, 1967; Marler and Hamilton, 1966; Hinde, 1966; Bateson and Hinde, 1976; Gould, 1982). Evolution of behavior usually means the study of animal behavior in relation to the environment and the identification of possible selective explanations for the distribution of behaviors within and among species.

Despite the emphasis on equilibrium analyses in ethology, a number of investigators have considered the historical and transformational implications of similarities and differences in behavior among organisms. Among the larger volumes on this topic, the books edited by Roe and Simpson (1958), Brown (1975), and Masterton et al. (1976) stand out as offering the most detailed treatment of ethology in a transformational framework. It is apparent from these books that a fundamental issue in historical ethology, the problem of homology recognition, has yet to be satisfactorily resolved. Concepts of homology and convergence are fundamental to the study of any historical pattern and to the testing of historical explanations in ethology. If ethology is to discern evolutionary patterns of behavior and if it is to contribute to generalizations about the process of evolution and to the recognition of patterns in nature, then it is important to critically evaluate conceptual approaches to the study of historical ethology.

In this chapter I attempt to clarify the issue of homology in ethology by focusing on two classes of criteria that have been proposed by ethologists as reliable indicators of behavioral homology: morphological and neural correlates of behavior. I begin by presenting an overview of historical analyses by ethologists and by reviewing proposals that homologous behaviors can be recognized by analyzing morphological and neural substrates of behavior. Of special importance for this paper are the following questions: (1) Are current research methods in ethology adequate for the analysis of historical patterns and transformational sequences of behavior? (2) Do a priori criteria often used by ethologists provide a useful method for determining behavioral homologies? (3) What are alternative approaches to the analysis of historical patterns and the recognition of behavioral homologies? and (4) How tightly coupled are historical patterns of morphological, neural, and behavioral features? Finally, I conclude with a case study that is used to illustrate one approach to the analysis of behavioral evolution. I know of no other such study where the evolution of a behavior has been studied comparatively (at the species level) within a clade at morphological, physiological, and ethological levels. This case study will serve as an example of an analysis of the historical relationship among behavioral, functional, and morphological characters, and will provide the basis for proposing hypotheses about the evolution of behavior and the evolutionary transformation of motor patterns.

HISTORICAL ANALYSIS IN ETHOLOGY

Common Research Approaches

Ethologists have several approaches to historical issues. Those issues of special interest in the context of a discussion of homology in ethology are:

1. The use of behavioral characters to produce phylogenies of organisms.
2. The use of ethological series as a basis for scenarios about the evolution of behavior.
3. The identification of homologous behaviors by the application of a priori criteria (i.e., criteria independent of phylogenetic patterns of character distribution). Of primary concern here is the concept that the structure of the nervous system producing the behaviors of interest and the morphology used in a behavior provide prima facie evidence for behavioral homology.

Before discussing these three issues, it is necessary to clarify the use of three terms that are important for this analysis. First, consider the terms "function" and "behavior." While some ethologists claim that their criteria for behavioral homology are drawn from the work of morphologists (e.g., Atz, 1970; Tinbergen, 1959; Eibl-Eibesfeldt, 1970), their use of the term "function" is very different from that of evolutionary morphologists. Cracraft (1981) noted the difference between function and behavior as concepts useful in studying historical patterns, and commented that "functions of structures generally have not been thought of as systematic characters, whereas ritualized behavior patterns have been utilized as characters for many years. . . . It is the behavior pattern, then, and not the function that constitutes a systematic character." Cracraft's use of the term "function" is consistent with that of most evolutionary morphologists: functions are the actions or uses of structures. But ethologists, in keeping with the dominance of equilibrium methods and the emphasis on environmental variables as determinants of behavior, have generally held that function "definitely means something quite different and specific: it means 'selective advantage'" (Hailman, 1976b, p. 184). Other ethologists have expressed similar views: "A function is a beneficial consequence of a behavior" (Bertram, 1976). "Within ethology, questions of function have been studied in two main ways: comparative studies of related species in which inter-species variations in a character are related to environmental variables . . . and detailed field studies involving either field experiments or the assessment of relations between variability in a trait and subsequent survival or reproductive success . . ." (Bateson and Hinde, 1976, p. 193). "A primary aim of primate socio-biology (or socio-ecology) is to explain variation in social behavior in terms of biological function. To do this, it is necessary to consider the possible consequences of differ-ences in behavior. . . . Some of these may be selectively advantageous, some neutral and some disadvantageous. Through this paper we refer to those in the first category as 'functions'" (Clutton-Brock and Har-vey, 1976, p. 195). The ethological definition of function is similar to what Bock and von Wahlert (1965) called "biological role." In this chapter I will use the term "function" in its morphological sense. This avoids the numerous difficulties associated with determining selection forces before one can meaningfully use the term "function." In addi-tion, the morphological definition will allow us to consider a method-ology for historical analysis in which behaviors, morphologies, *and* functions can be used both as systematic characters and as elements of an overall transformational pattern with general implications about the evolution of behavior.

Secondly, consider terms that could be applied to features of organ-isms that are mistakenly believed to be homologous. Patterson (1982, p. 45) distinguishes five possible terms that represent nonhomologous (homoplasous) features: parallelism, convergence, analogy, mimicry, and chance. To this list we might add learned similarities. I will include the last four terms of Patterson's list and learned behavioral similarities under the label "convergence." Thus, my use of the word "conver-gent" includes nonhomologous features that may be the result of sev-eral processes.

1. Behavioral characters and phylogeny. Perhaps the most common avenue of research into historical problems in ethology is the use of behavior to clarify the relationships of organisms. Since the early days of ethology, behaviors or components of a behavior have been used to infer phylogenetic relationships (Whitman, 1899; Lorenz, 1950; Tin-bergen, 1959; Miller and Jearld, 1982; Miller and Robison, 1974; Mayr, 1958; Eberhard, 1982; Greene, 1977). The procedures for such research are usually grouped under the rubric "comparative method" (Hailman, 1976a), and are often held to be similar to the methods used by morphologists to study phylogenetic relationships. As outlined by Hinde and Tinbergen (1958), comparative analyses of behavior usually begin with the examination of several species and the recognition of behavioral similarities among the species. Then, either the patterns of behavioral similarity can be compared to a currently accepted phylog-eny (Greene and Burghardt, 1978; also see Mayr, 1958) to see how similar the pattern of relationships based on behavioral characters is to that based on morphological evidence, or the behavioral similarities themselves can be used to construct a phylogeny, as in the work of Lorenz (1941). Two important concerns have often been expressed by ethologists in relation to the process of using behavioral characters in phylogenetic analysis. The first is the problem of the units of behavior. What are the appropriate components of an organism's behavioral repertoire to use in a comparative analysis? Many workers have suggested that "fixed" action patterns be used in phylogenetic analysis because of their relative stereotypy and their consequent ease of identification (e.g., Bateson and Hinde, 1976). The second issue is the variability of behavior. If behavioral units appropriate for comparison cannot be reliably identified and if it is not possible to identify homologous similarities (an issue considered in detail below), then behavior would appear to have little value in historical analysis. This has led several investigators to conclude that behaviors have no utility whatsoever in the analysis of patterns of relationship (Aronson, 1981; Atz, 1970).

2. Ethological series. A second method used by ethologists to study the evolution of behavior is similar to the morphological series approach of morphologists (Lauder, 1981). If an investigator wishes to analyze the evolution of a complex behavior, for example, a comparative approach is taken in which taxa exhibiting similar but less complex behaviors are arranged in a series from the least complex to the most complex. This series is often taken as representative of a possible evolutionary pathway leading to the complex behavior, and interpretations of selection forces, behavioral transformations, and causative environmental factors are based on this series. A good example of this procedure is the analysis of Kessel (1955), who studied courtship behavior of balloon flies.

This approach has several pitfalls, not the least of which is the probable lack of congruence between a behavioral or morphological series and the pattern of transformation derived from a corroborated phylogenetic hypothesis of the relationships of the clade under study (Lauder, 1981). Behavioral or morphological series are artificial constructs of equilibrium points and are nonhistorical in nature.

3. Criteria for homology. The final and most important issue to consider in this examination of current approaches to historical patterns in ethology is the problem of recognizing behavioral homologies. Many criteria have been proposed by ethologists for determining which behaviors in a comparative analysis are homologous, and many workers have discussed extensively the value of these criteria. Virtually without exception, however, these discussions have had one premise: that homologous behaviors can be recognized by the application of criteria that are largely independent of phylogenetic patterns of character distribution. I term such criteria *a priori criteria for homology* because no reference is made to the phylogenetic distribution of other characters possessed by the clade. Three such a priori criteria have been of particular importance in ethology:

1. Two behaviors are homologous if the neural and neuromuscular control systems of the two behaviors are similar. I will refer to this as the *neural criterion*.

2. Two behaviors are homologous if the gross morphological features used in the behavior are homologous. I will refer to this as the *morphological criterion*.

3. Two behaviors are likely to be analogous if the biological role (function in ethological terminology) of the two behaviors is similar, and/ or if the two behaviors could have been subjected to similar selection pressures. I refer to this as the *selection criterion*.

First, consider the neural criterion, which is perhaps the most commonly proposed a priori criterion for determining homologies. Baerends (1958, p. 409) has stated one version of this criterion clearly: "Our considerations lead to the conclusion that in comparative ethology it is most essential for homology that the patterns of muscle contraction should be largely identical." Other investigators have been equally impressed with the capability of the central nervous system of providing a reliable guide to behavioral homology, as the following quotations demonstrate:

> Thus, uncovering a behavioral process which, in spite of superficial modifications, is shown to depend on homologous neural structures provides a valid criterion useful in a taxonomy of behavior—and valid criteria for classification are not abundant in the behavioral sciences. (Pribram 1958, p. 142).

> The study of the evolution of the capacities to acquire new behaviors leads inevitably into the central nervous system as the locus of homologous structural correlates through which homologous behavioral capacities can be identified. (Hodos, 1976, p. 162).

> Behaviors associated with brain structures that have a common genealogical history are homologous, whether or not the behaviors are of the same type or serve the same function to the animal.
> (Hodos, 1976, p. 163).

Two concepts would appear to underlie application of the neural criterion: the idea that the central nervous system and the motor patterns it produces exhibit more conservative evolution than do behaviors, and the idea that there is a tight coupling between neural structure and behavior. In the absence of evidence supporting these ideas, it would be difficult to suggest that the nervous system is the map that allows us to determine behavioral homologies. We need an explicit methodology for evaluating the linkage between neural and behavioral evolution.

There are several difficulties that arise with the application of the neural criterion, ignoring the fact that we know relatively little about the neural control of behavior, although the field of neuroethology is showing rapid progress (Hailman, 1976b). A simple theoretical example will introduce the discussion.

Consider the approach depicted in Figure 1-1. A clade of ten species is studied (A to I, with an outgroup species O) and a certain behavior is found to be present in five of these species (B, E, G, H, I, boxes in upper panel). This behavior is identical in these species. An analysis of the morphological and neural bases of this behavior reveals that species G, H, and I all show morphological and neural correlates of the behavior, but that species B and E possess only one correlate. Using an accepted cladogram (Figure 1-1, middle panel) as a basis, we can map the distributions of the behavioral, morphological, and neural novelties onto the cladogram to provide a basis for historical inferences about the evolution of behavior. In this case (Figure 1-1, lower panel), we see that the behavioral pattern shown by species B is not homologous with that of species E, and that the behavior is homologous within the monophyletic clade formed by species G, H, and I.

Figure 1-2 illustrates two possible outcomes of a theoretical study in the evolution of behavior relevant to our consideration of the neural criterion. Imagine that ten species of shrimp have been studied (A to I, with an outgroup species O) and that six species are found to display a particular escape behavior. High-speed films of this behavior demonstrate that the behavior is identical in all six species. The mapping of the occurrence of this behavior onto the cladogram (accepted as true on the basis of other characters) demonstrates that the behavior is homologous in species A, B, and C and in species G, H, and I, but not between these two groups. Thus, the behavior has evolved at least twice in this clade of shrimp.

Further neurobiological study reveals that in some species one neural circuit mediates this behavior, while in other species a completely different circuit composed of different motoneurons innervating different peripheral muscles controls the behavior. If the results of this study are as shown in the upper panel of Figure 1-2, then we may conclude that the independent evolution of similar escape behaviors in the two groups of shrimp was accompanied by the evolution of two different neural control systems for the behaviors. In this case, the patterns of variation in behavior and the neurobiology are historically congruent.

However, if the results of the study produce the pattern shown in the lower panel of Figure 1-2, then an entirely different conclusion may be drawn. In this case, a homologous behavior (the escape behavior in

TAXA / FEATURES	A	B	C	D	E	F	G	H	I	O
MORPH	●		●		●		●	●	●	
NEURAL	○	○		○			○	○	○	
BEHAV		□			□		□	□	□	

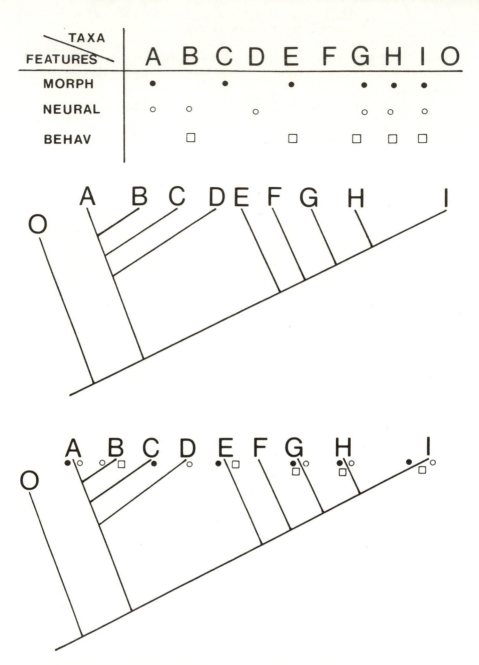

FIG. 1-1   A simple theoretical example to illustrate one procedure for studying the evolution of behavior and the neural and morphological bases of behavior. In this example, species A to I are studied (O is an outgroup clade) and several of the species are found to possess a particular behavior pattern indicated by the boxes in the upper panel. The behavior is identical in each of these species. Study of the morphological and neural patterns in these species shows that some species share a particular specialization indicated by the solid and open circles respectively. Using an accepted cladogram as a basis (middle

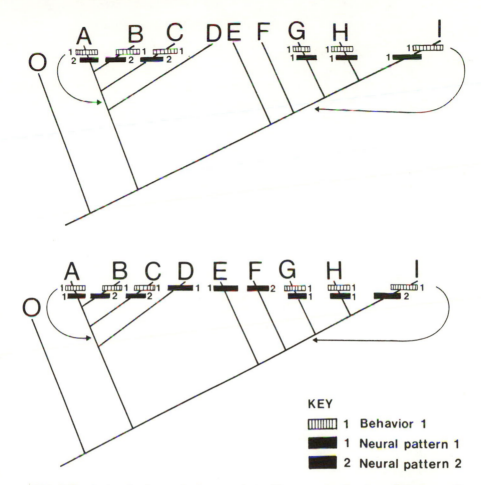

KEY

▦ 1 Behavior 1
■ 1 Neural pattern 1
■ 2 Neural pattern 2

FIG. 1-2 A simple theoretical example to illustrate application of the neural criterion for determining behavioral homologies. The two panels illustrate two types of mapping of behavioral and neural patterns we might expect if there is a tight historical coupling between neural and behavioral features (upper panel) and a loose coupling (lower panel). The distribution of two neural patterns is shown in relation to one behavior. Note that in both panels the behavior is homologous in taxa A, B, and C, as well as in taxa G, H, and I, but that the behavior is not primitive for the clade as a whole and thus is not homologous in these two groups. The cladogram is accepted as being correct; of course, the behavioral and neural characters could be used to modify the cladogram for subsequent analyses. Arrows indicate the level where the behaviors are inferred to have evolved. Also note that the distribution of characters on the lower panel shows that homologous behaviors may have nonhomologous neural substrates.

panel), the morphological, neural, and behavioral novelties may be mapped onto the cladogram (bottom panel). Inferences based on this pattern provide insight into the evolution of behavior.

species A and B) is mediated by nonhomologous neural systems. In addition, there is no clear relationship between the phylogenetic distribution of escape behavior and the type of neural circuit controlling the behavior. Here, the neural criterion fails.

The most important point is that the success or failure of the neural criterion depends entirely on the phylogenetic distribution of the neural components of a behavior in relationship to the phylogenetic distribution of the behavior itself. In some systems or types of behavior, there may be a tight coupling between neural and behavioral evolution, while in other cases there may be little relationship between behavioral and neural evolutionary novelties.

There are four specific reasons, grounded in recent findings in comparative neurobiology and neuroethology, why I believe that the neural criterion is not adequate as an indicator of behavioral homology. First, neural connections and structures can exhibit convergent evolution just like any morphological feature. There is no evidence to suggest an overall phylogenetic conservatism of neural connections or centers that could support use of the nervous system as an arbiter of behavioral homology. For example, Northcutt (1981) notes that pallio-spinal and cervical-spinal pathways apparently evolved independently in galeomorph sharks, urodele amphibians, birds, and mammals. He summarizes one aspect of recent progress in neuroanatomy: "Perhaps we are discovering that active vertebrate predators have independently and repeatedly evolved complex and very similar nervous systems to solve a similar array of environmental problems." If features of nervous systems are just as likely to be convergent as behaviors or skeletal features are, then what basis is there to assign priority to neural information in determining homologies?

The second objection to the neural criterion stems from constraints on the design of neural circuits. To take one example, the types of possible circuits that produce patterned behavior are very few (four or five basic types). Each of the basic neural electrical circuits has a particular functional significance, and animals that produce patterned movements of a certain type tend to have those types of controlling circuits (see Bullock, 1961; Delcomyn, 1980). A similar argument has been made by Ulinski (1984) for constraints in the design of vertebrate sensory systems. In short, the range of possible engineering designs of pattern generators is limited, and convergence of neural control in unrelated animals that produce grossly similar patterned behaviors should be expected. Evidence of a similar neural circuit may have nothing to do with homologous behaviors because of the constraints on circuit design.

Thirdly, one neural circuit can mediate several different behaviors depending on the pattern of nerve impulses in the circuit. Are we to consider that all of the different behaviors controlled by a single circuit are homologous to one another because they have a common neural basis? Croll and Davis (1981) provide an example of a situation in which a variety of motor *programs* are used to control the same motor *system* in *Pleurobranchia*, and similar findings were obtained by Ayers and Davis (1977) and Ayers and Clarac (1978) in their studies of lobster *(Homarus)* walking. Similar neural circuits and similar peripheral morphologies (arrangements of muscles and bones) are no guarantee that the behaviors produced by these systems will be similar. The converse of this third objection to the neural criterion also holds: very similar behaviors can be elicited by different patterns of neural stimulation. A conclusion of Hoyle's (1976) research on the neurophysiology of insect locomotion, in which the pattern of motor output to locomotor muscles during normal locomotion was very variable, was that "widely different patterns underlie even quite similar movements."

The fourth and final objection to the neural criterion is that homologous neural circuits can mediate different behaviors because of changes in the peripheral morphology. If muscles change their origins and insertions in one group of species and the neuromuscular output to those muscles remains the same as in other species in the clade, then the patterned movements produced by the bones will be different in the clade with the new muscle arrangement, even though the neural substrate may be homologous. An example of this can be found in the body movements of decapods used in locomotion. Paul (1981a, 1981b) proposed homologies between the motoneurons, muscles, and motor output to the muscles in the tails of two unrelated decapod families. Yet the movements produced by these homologous systems are very different in the two families.

These four arguments based on comparative neurobiology have been adduced to demonstrate that there is little support for the neural criterion in the literature of neurobiology. Furthermore, the simple example of Figure 1-2 illustrates that a phylogenetic approach (discussed in more detail in the next section below) indicates that behaviors may be homologous even though the neural substrates are not. There is no reason to presume a one-to-one mapping between the nervous system and behavior; each is capable of divergent evolutionary patterns.

The second a priori criterion for determining behavioral homologies mentioned above is the morphological criterion. This is similar to the neural criterion in that both rely on an underlying structural substrate to a behavior as the arbiter of homology. This criterion is perhaps the most widely cited, and the following quotations give a sample of ethological opinion on this issue:

> The extent to which behavior can be homologized is directly correlated with the degree to which it can be conceived or abstracted in morphological terms. (Atz, 1970, p. 68).

It is manifestly impossible for nonhomologous structures to have homologous functions. (Atz, 1970, p. 60).

Behavioral homology has been defined here in the context of specific structural entities that are homologous. . . . Behaviors associated with . . . two homologous structures would therefore be homologous no matter how different the function of the two behaviors were (Hodos, 1976, pp. 160–161).

The concept of behavioral homology is totally dependent on the concept of structural homology. . . . If evidence comes to light indicating that structures previously thought to be homologous are not homologous, then any behaviors associated with these structures must also be redesignated non-homologous (Hodos, 1976, p. 165).

Functions, considered as abstractions and without consideration for the structures that perform these functions, should not be spoken of as homologous. (Haas and Simpson, 1946, p. 323).

Divergence and lack of homological behavior between insects and vertebrates are again illustrated, for the external skeleton-internal muscle apparatus of an insect obviously had a different origin from the internal skeleton-external muscle apparatus of a vertebrate (Simpson, 1958, p. 509).

The morphological criterion may be subjected to a similar analytical procedure to that used for the neural criterion (Figures 1-1 and 1-2). Mapping behavioral and morphological (structural) novelties onto a cladogram will allow us to determine if there is a tight historical linkage between the two. If, as in Figure 1-1, homologous behaviors can have nonhomologous morphological substrates, then the morphological criterion will be of little value in determining behavioral homology. Because there is so little data bearing on the problem of congruence between morphological and behavioral novelties, excessive reliance on morphology as an arbiter of behavioral homology may lead to erroneous conclusions.

An important consideration in an analysis of the morphological criterion is the problem of the homology of functions (use of structures). Under what circumstances can functions be considered to be homologous? Haas and Simpson (quoted above) felt that unless there was a definable morphological substrate, functions could not be homologous. If we choose to identify convergences by relying on a criterion of similar functions (as "the wing of a bat and the wing of an insect are convergent because they have similar functions but different structures") then it will be difficult to recognize homologous functions. I suggest that functions can be homologous, can be used as systematic characters, and can be recognized as such on the basis of their phylogenetic distribution (examples will be given in the case study).

One example of the confusion that can result from the application of the morphological criterion to the homology of functions is found in the following statement: "The insertion of food into the mouth by a man and a monkey would be both homologous (because the hands of monkeys and humans are derived from hands of their common ancestors) and analogous (because the behaviors serve the same purpose)" (Hodos, 1976, p. 160). One might question the utility of a criterion that leads to the determination of a behavior as both homologous and convergent.

The third a priori criterion for the recognition of behavioral homologies is the likelihood that the behaviors under consideration could be subject to selection and thus could be convergent. This criterion relies on the argument that behaviors that are subject to strong selective forces are more likely to be convergent.

Illustrative remarks in support of the selection criterion follow:

> A knowledge of the selective factors exerting pressure on the animal will, therefore, help to judge how far in a certain case the possibility of convergency has to be taken into consideration (Baerends, 1958, p. 409).

> Behavior is subject to particularly strong selection. . . . Strong selection pressures, especially those associated with rigorous environments, tend to result in convergences (Atz, 1970, p. 64).

> Parallelisms and analogies are particularly common in all types of behavior that are strictly functional, such as food getting or locomotion (Mayr, 1958, p. 351).

The selection criterion may be criticized by noting the difficulty of demonstrating that any feature is under selection pressure. Indeed this criterion assumes *directional* selection, and that the selection pressures on behaviors in the present habitat were those that shaped the behaviors in the past. Cracraft (1981) has criticized morphologists for their axiomatic acceptance of concepts of selection and their failure to undertake research programs that *test* the role of selection in producing morphological patterns. The same criticism can be applied to the selection criterion of homology recognition: the use of selection is axiomatic, not deductive. There is to my knowledge no way to determine selection pressures (magnitude and direction) and their effect on behavior by simply examining the behavior of a few individuals. Knowledge of intrapopulation variability in relation to fitness is necessary (Lande and Arnold, 1983), and a detailed, specific research program for studying selection has been outlined by Arnold and Wade (1984a, 1984b). I conclude that the axiomatic application of the selection criterion is unlikely to contribute to decisions about the likelihood of homology and convergence of behavior, and that broad statements about the significance of selection for the study of behavior only mask our ignorance about the nature of variation in behavior, its genetic basis, and its relation to fitness.

HISTORICAL ETHOLOGY

Since the early 1970s, morphologists and systematists have made major advances in our understanding of research methodologies and concepts in comparative and historical biology. Unfortunately, ethologists have generally not been aware of this progress, and as a result most recent discussions of homology and convergence have not utilized recent terminology or benefited from modern conceptual clarifications. The approach taken here will be similar to that of recent discussions of homology by comparative biologists (see Eldredge and Cracraft, 1980; Wiley, 1981; Cracraft, 1981; Nelson, 1970; Patterson, 1980, 1982). My aim is to illustrate the conceptual basis of one approach to historical ethology: the comparative phylogenetic approach. This theoretical discussion will serve as the basis for the specific example discussed in the case study below.

A simplified scheme for phylogenetic analysis may be outlined as follows. First, some preliminary decision must be made about which taxa are to be included in the study and the limits of those taxa. Next, similarities among the taxa are recognized. These similarities are provisional hypotheses of homology (Cracraft, 1981). Third, the similarities are clustered into nested sets defining a hierarchy of taxa. In general the clustering procedure uses a set of working rules that minimizes the number of steps used to produce the hierarchy. For analyses with many taxa and many similarities (characters), computer programs are available that produce one or more branching diagrams based on a variety of different assumptions. Such a hierarchy can be expressed as a branching diagram (e.g., Figures 1-1 and 1-2). Invariably, some of the proposed similarities from the second step are found to indicate a relationship between two taxa that is in conflict with the relationships indicated by other similarities. Those conflicting characters are *convergences*, as they are in conflict with the hierarchical pattern corroborated by the majority of other characters. As Patterson (1980, p. 236) notes, "In other words, these characters fail the principal test of homology, congruence with other characters."

The character conflicts in the branching diagram could have been produced by a variety of processes such as learning, convergent evolution, mimicry, and chance. Additional research would be needed to evaluate explanations for each character conflict. Such additional investigation of specific characters is often undertaken if no clear branching diagram emerges from the clustering of characters into nested sets. Thus, if an attempt to generate a corroborated branching diagram results in as many (or more) characters conflicting with the diagram as supporting it, then individual character conflicts may be investigated in an attempt to better resolve the pattern of character distribution.

The most widely discussed source of additional information on characters is ontogeny. The pattern of development may provide data that resolve conflicts in the branching diagram by demonstrating that two characters initially thought to be similar (and thus providing evidence of relationship between two species) are in fact different.

Key results of this method of phylogenetic analysis are that *homologous similarities are those that define monophyletic (natural) groups* (Patterson, 1982), and that we recognize similarities as homologies only as a *consequence* of examining the distribution of other characters. Thus, homologies are recognized a posteriori, and are a consequence of accepting a particular phylogeny as being an accurate depiction of the pattern of ancestry and descent. To use the example from Figure 1-2, it is unparsimonious to assume that the escape behavior is homologous between species C and G, as this would imply that taxa D to F had lost this feature. If we accept the evidence from other characters that the branching diagram shown in Figure 1-2 is the corroborated hypothesis of phylogenetic relationship, then escape behavior has evolved twice. Thus, as discussed by Patterson (1982), homologies characterize natural groups and such groups are individuals in the sense of Hull (1978) and Ghiselin (1974); individuals are coherent spatiotemporal entities that participate in natural processes (Wiley, 1981). Analogies, on the other hand, characterize *classes* (Ghiselin, 1984) of taxa, groupings that do not participate in natural processes and are spatiotemporally unrestricted. Paraphyletic and polyphyletic groups are classes. Ethologists unfamiliar with these concepts in phylogenetic analysis are referred to the general texts by Wiley (1981) and Eldredge and Cracraft (1980), and the excellent discussions of homology in Patterson (1980, 1982).

To take one final example from the field of comparative neuroethology, consider the distribution of species of fishes that can use electroreception to obtain information from the environment. Electroreception appears to be a sensory modality that is primitive for vertebrates (Bullock et al., 1983). However, within the teleost fishes (which primitively lack an electric sense), electroreception has reevolved several times. Even though the mormyrid fishes and the gymnotid fishes both possess similar types of ampullary receptors, have afferents that enter the brain through the anterior lateral line nerve, and have similar specializations of the medial nucleus, the evidence for convergent evolution of this sensory modality is convincing because of the large number of other similarities that corroborate a phylogenetic hypothesis different from one based on the concept that the electroreceptive systems are homologous.

While this may seem a trivial and obvious point, the implications of the congruence criterion for homology recognition have not been appreciated by ethologists, nor, for the most part, have the implications been incorporated into comparative ethological research programs. Some investigators have defended the view that methods and procedures of phylogenetic analysis are not applicable to the study of behavior: "methods—such as phylogenetic trees, cladograms, and homologies—used for the study of phenomena at one level (say morphology) are generally not applicable to higher-level phenomena (say behavior)," and "one should be aware of the pitfalls in trying to build phylogenetic trees of behavior or to establish behavioral homologies across large phyletic gaps by using morphological methods" (Aronson, 1981, p. 37). Atz (1970) has expressed similar reservations based on the difficulty of recognizing behavioral homologies by a priori criteria.

If the primary test of homology is congruence with other characters, then there is no reason to assume prior to a phylogenetic analysis that one particular class of characters (such as behaviors) will not be useful indicators of phylogeny and will be expected to show a great deal of convergence. Furthermore, there is no reason to rule out the analysis of functions (uses of structures) as possible characters for understanding transformations of biological designs and behaviors. To my knowledge, no study has considered patterns of distribution of function, behavior, and morphology within a historical context, and no attempt has been made to explicitly define the transformational relationship among these three types of biological features. We will never know if behaviors or functions are too variable or have low utility for studying evolutionary transformations unless research programs explicitly address the question of historical congruence among behaviors, functions, and morphological patterns in an attempt to identify general features of biological transformations.

A CASE STUDY IN THE EVOLUTION OF BEHAVIOR

Goals and Research Approach

The purpose of presenting this case study is to provide a detailed example of a phylogenetic and historical approach to the analysis of behavioral, morphological, and functional patterns. The approach taken will be a posteriori in that homologous characters will be identified on the basis of their congruence with a phylogenetic hypothesis. This case study addresses a general problem in ethology: the relationship among the transformation of form, function, and behavior.

The subject of the case study is the feeding behavior of sunfishes (family Centrarchidae), an endemic North American family of 32 species. Within this family two species exhibit a derived feeding behavior and eat snails; the snails are crushed in the pharyngeal area of the mouth before being swallowed. Most other sunfish species have a relatively diverse diet (Keast, 1978a, 1978b; Savitz, 1981). The two species that do eat snails, the pumpkinseed sunfish *(Lepomis gibbosus)* and the redear *(L. microlophus)*, have specialized feeding behavior in that the snail shells are crushed in the pharynx and then the shell is separated from the body and ejected from the mouth before the body itself is swallowed (Lauder, 1983a, 1983b). The snail is not swallowed whole.

An analysis of the evolution of this snail-crushing behavior is particularly suitable for addressing the general questions raised above because (1) the behavior is specialized and is not widespread within the family, (2) there are morphological specializations in the feeding mechanism of species that regularly exhibit the crushing behavior, (3) the physiological basis of the crushing behavior can be studied experimentally by recording electrical activity from the muscles that are involved in crushing the snails (thus the motor patterns used in the behavior can be identified), and (4) the crushing behavior has previously been identified as being composed of repetitive crushing phases (Lauder, 1983b). As Selverston (1976, p. 82) asserts, "If we are to understand how nervous systems generate behavioral activities, cyclically repeating motor patterns are a good place to start. By their very nature, they allow repeated study of the basic mechanisms involved." Other authors have emphasized the importance of studying the underlying physiological and motor bases of behaviors (e.g., the papers in Fentress, 1976, by Bullock, Kennedy, and Hoyle; Barlow, 1977; Greene and Burghardt, 1978; Delcomyn, 1980), but few studies are available in which such comparative physiological patterns are available for taxa in which concomitant behavioral, morphological, and phylogenetic analyses have been conducted.

In this case study, a particular phylogenetic hypothesis is taken as a starting point from which to consider the patterns of behavioral, morphological, and physiological features. This phylogenetic hypothesis, derived from work currently in progress, indicates the corroborated relationships of some of the centrarchid species, and does *not* rely on characters discussed here for evidence supporting the phylogeny. Thus, the characters that are discussed in this paper were not used to make the phylogeny initially. Given this pattern of genealogical relationship as a starting point, the characters of interest in this analysis are compared to the phylogeny. Snail-crushing behavior, morphological specializations in the feeding mechanism, and motor patterns are all mapped onto the phylogeny to determine if patterns of evolution in these characters are congruent. This procedure allows us to assess the neural and morphological criteria for homology. The neural criterion would be corroborated, for example, if the phylogenetic distribution of the motor patterns producing the snail-crushing behavior was congruent with the phylogenetic distribution of the behavior itself.

Materials and Methods

Individuals of all genera in the family Centrarchidae were studied morphologically, and most species within the largest genus, *Lepomis*, have also been examined. Dissections of the head were used to ascertain the basic design of the feeding mechanism. Muscle orientations were examined in most species to determine the range of variation in the clade, and muscle size was studied by measuring the physiological cross-section of the muscles in the pharynx. Physiological cross-sections (Gans and Bock, 1965) provide a relative measure of the force a muscle is capable of generating, and this measure takes into account differences between muscles in fiber arrangement and length. Details of the measurements and techniques may be found in Lauder (1983b). The size of teeth on the pharyngeal bones was quantified by digitizing the areas of teeth on the pharyngeal jaws of nine species.

Information on the trophic biology of sunfish species was obtained from the literature and from laboratory observations. Only two species, *Lepomis gibbosus* and *L. microlophus*, would regularly eat snails in the laboratory, and these two species also are the chief natural molluscivores within the Centrarchidae (Savitz, 1981; Gosline, 1985). Individuals of one other species of centrarchid, *Lepomis cyanellus* (green sunfishes), would eat snails in the laboratory. Other species such as the bass *(Micropterus)* or the bluegill sunfish *(Lepomis macrochirus)* refused to eat snails.

In order to determine the motor patterns involved in feeding behavior, electrical recordings were made from the pharyngeal muscles in unrestrained, active fishes. Electromyograms were recorded using fine-wire bipolar electrodes implanted percutaneously into pharyngeal muscles in anesthetized fish. When the fishes recovered from anesthesia, the electrode leads (see Lauder, 1983b, for details) were attached to Grass P511J amplifiers and signals were recorded on FM tape for subsequent analysis. Nine species in the Centrarchidae have been studied experimentally to date: *Ambloplites rupestris, Pomoxis annularis, P. nigromaculatus, Micropterus salmoides, Lepomis cyanellus, L. gulosus, L. macrochirus, L. gibbosus,* and *L. microlophus*. In addition, electromyographic recordings were made from hybrids between *L. microlophus* and *L. cyanellus*. Three types of food were used: pieces of earthworm *(Lumbricus,* 1 to 8 cm long), snails *(Helisoma,* 2 mm to 1 cm in diameter), and small minnows *(Pimephales,* 3 to 6 cm total length).

Results

Figure 1-3 shows a schematic view of the feeding apparatus of a centrarchid fish. The oral jaw apparatus is located at the front of the buccal cavity (BC, OJA) and is used during initial prey acquisition by the suction feeding mechanism characteristic of teleost fishes (Lauder, 1982). Once the prey has been captured and is located in the buccal cavity, water flow through the mouth is used to move the prey posteriorly to the pharyngeal jaw apparatus (PJA). The pharyngeal jaw bones are used to move prey into the esophagus (ES) and to manipulate prey prior to swallowing. Figure 1-3 shows the position of a snail between the upper

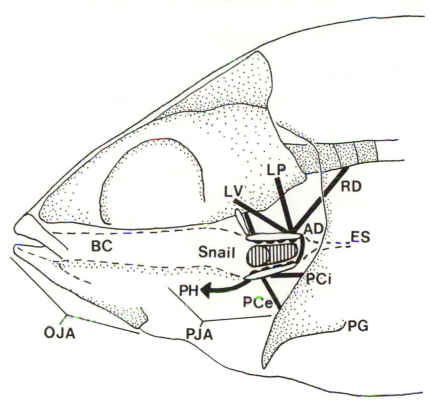

FIG. 1-3 Schematic view of the head of a centrarchid sunfish to show the location of the major morphological units associated with feeding behavior. Fishes possess two major sets of jaws: an oral jaw apparatus (OJA) and a pharyngeal jaw apparatus (PJA). A snail is shown between the pharyngeal jaws in the position used by the fish for crushing. Abbreviations: AD, fifth branchial adductor muscle; BC, buccal cavity; ES, outline of the esophagus; LP, levator posterior muscle; LV, external and internal branchial levator muscles; OJA, oral jaw apparatus; PCe, pharyngocleithralis externus muscle; PCi, pharyngocleithralis internus muscle; PG, pectoral girdle; PH, pharyngohyoideus muscle; PJA, pharyngeal jaw apparatus; RD, retractor dorsalis muscle.

and lower pharyngeal jaw bones when the shell is being crushed (as determined by x-ray cinematography, Lauder, 1983b). The upper and lower pharyngeal jaws are connected by a short muscle, the fifth branchial adductor (AD). Many other muscles (some of which are shown in Figure 1-3) attach the pharyngeal jaws to the skull and surrounding bony elements such as the pectoral girdle (PG). This basic musculoskeletal design is common to all centrarchid fishes and does not vary within the species studied. Snail-eating sunfishes thus possess the same basic design as other trophic types in the family, and no new muscles, bones, or muscle origins or insertions are present.

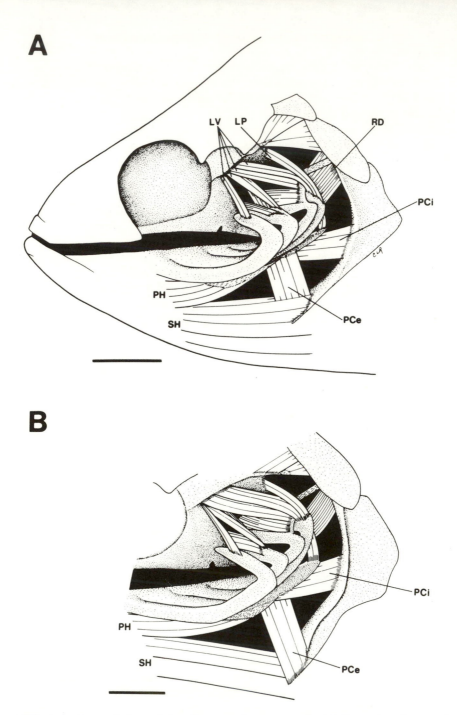

FIG. 1-4 Lateral view of the pharyngeal apparatus in (**A**) *Lepomis gulosus* and (**B**) *Acantharchus pomotis* to illustrate the anatomical organization of this region. Comparative illustrations of other centrarchid fishes are presented in Lauder (1983b). Scale = 5 mm. Abbreviations: SH, sternohyoideus muscle; others follow those in Figure 1-3.

The musculature of the pharyngeal region is shown in more detail for two species in Figure 1-4. The upper pharyngeal jaw on each side is attached to the skull by a series of branchial levator muscles (LV) and to the vertebral column posteriorly by the retractor dorsalis muscle (RD). Ventrally, the lower pharyngeal jaw on each side is attached to the pectoral girdle by the pharyngocleithralis internus and externus muscles, and to the hyoid anteriorly by the pharyngohyoideus (PCi, PCe, PH).

Examination of the physiological cross-sections of the pharyngeal muscles in four species (Table 1-1) shows that the two snail-eating species share two characters: hypertrophy of the pharyngohyoideus (PH) and levator posterior (LP) muscles. None of the other muscles in the pharynx of these two species is significantly enlarged relative to pharyngeal muscles in other species.

The pharyngeal jaw bones of several representative species of centrarchids are illustrated in Figure 1-5. The lower pharyngeal jaw is composed of a pair of bones that attach anteriorly via ligaments to the branchial apparatus. The upper pharyngeal jaws consist of several bones (pharyngobranchials and epibranchials) that are closely articulated with each other and bear enlarged toothplates. *Lepomis microlophus* and *L. gibbosus* have robust lower pharyngeal jaws in comparison to the other species, and have larger and more rounded teeth (Figure 1-5; Table 1-2). In both of these species, the upper pharyngeal jaw teeth are hypertrophied with respect to the lower jaw teeth (Table 1-2).

Two basic types of motor patterns in the pharyngeal muscles were discerned in the species studied electromyographically. First, during the swallowing of fishes and worms, all species except one (the redear, *L. microlophus*) display a rhythmic pattern of muscle activity that may continue for up to a minute as the prey is moved from the pharyngeal area into the esophagus. Examples of this rhythmic swallowing pattern are shown in Figure 1-6; more quantitative representations of this sequence of muscle activity are given elsewhere (Lauder, 1983b). Note that there is a considerable difference between the activity patterns of different pharyngeal muscles (e.g., the PCe and PCi in Figure 1-6B),

TABLE 1-1 Branchial muscle physiological cross-sections (mm^2) in centrarchid sunfishes. Data are given for eight muscles in four species. See text for discussion. Muscle abbreviations: AD5, fifth branchial adductor; LE1, levator externus one; LE3/4, levatores externi three and four; LP, levator posterior; PCe and PCi, pharyngocleithralis internus and externus; PH, pharyngohyoideus; RD, retractor dorsalis.

	RD	AD5	PH	PCi	PCe	LE1	LE3/4	LP
Lepomis macrochirus	5.2	1.8	2.0	3.5	5.8	1.0	2.8	1.0
L. cyanellus	9.3	4.1	1.9	3.5	4.3	0.9	3.4	1.7
L. microlophus	3.2	3.5	7.3	5.1	5.5	1.2	3.9	16.7
L. gibbosus	6.9	6.3	6.7	6.7	10.7	2.1	3.6	11.3

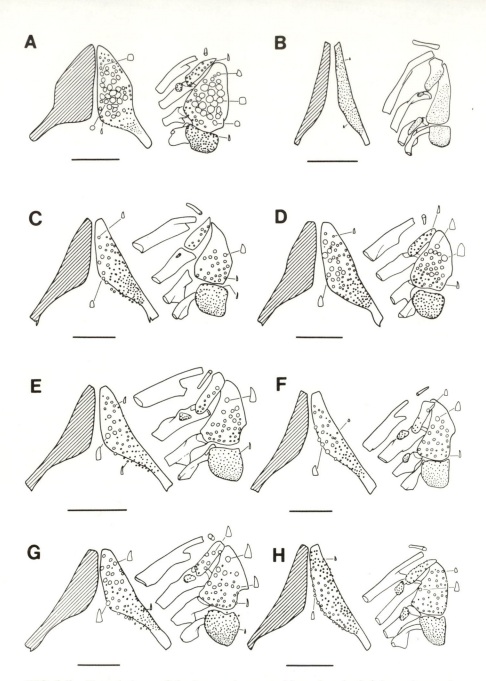

FIG. 1-5 Dorsal views of the lower pharyngeal jaws (on the left in each panel, with the left jaw shaded to emphasize shape differences), and a ventral view of the upper pharyngeal elements of the right side of the head (on the right in each panel). Representative teeth are shown in side view. Note the differences between the structure of the pharyngeal jaws in *Lepomis microlophus* and *L. gibbosus* in comparison to the other species. **A,** *Lepomis microlophus;* **B,** *Pomoxis nigromaculatus;* **C,** *L. punctatus;* **D,** *L. gibbosus;* **E,** *L. auritus;* **F,** *L. cyanellus;* **G,** *L. megalotis;* **H,** *Ambloplites rupestris.*

TABLE 1-2 Pharyngeal tooth areas in sunfishes (mm²) on the upper pharyngeal jaws (UPJ) and lower pharyngeal jaws (LPJ).

	UPJ	LPJ
Ambloplites rupestris	0.008	0.006
Lepomis auritus	0.012	0.016
L. megalotis	0.028	0.021
L. microlophus	0.066	0.046
L. punctatus	0.013	0.015
L. cyanellus	0.017	0.013
L. gibbosus	0.071	0.029
L. macrochirus	0.011	0.014
L. gulosus	0.009	0.004

and that the muscle activity shown represents only a small portion of a much longer duration sequence of this rhythmic activity.

The second basic type of motor pattern was observed when snail shells were crushed between the pharyngeal jaws. During snail crushing, nearly all pharyngeal muscles were electrically active simultaneously, and no rhythmic alternating pattern was observed (Figure 1-7B and C). In some experiments a hydrophone was placed in the aquarium to determine when in the course of muscle activity the snail shell cracked (see Figure 1-7B). The first sign of shell failure invariably occurred at the end of the burst of muscle activity as illustrated in the bottom trace of Figure 1-7B, and successive bursts of muscle activity produced less sound until the shell was completely crushed. After being crushed, the shell is separated from the snail by a different motor pattern from that used in crushing, and the shell fragments are ejected from the mouth cavity. The snail is then swallowed.

Statistical analyses reveal that the motor pattern used in snail crushing is less variable than that used in swallowing fish or worms (Lauder, 1983b). Muscle activity bursts are shorter and pharyngeal muscles are more similar to each other in the crushing motor pattern than in the swallowing pattern used for fish and worms. One striking result of this comparative analysis of the motor patterns used in processing food in the pharynx is that one species *(Lepomis microlophus)* uses the crushing pattern for *all* types of prey, even fish and worms (Lauder, 1983a). The other species that eat snails (*L. gibbosus* and *L. cyanellus*) use the crushing motor pattern when feeding on snails, but use the rhythmic swallowing pattern when feeding on fish and worms (compare A and C in Figure 1-7). These species, in contrast to *L. microlophus*, modulate the motor pattern to the pharyngeal muscles depending on the type of prey they are processing. On the other hand, the redear sunfish uses the same basic motor pattern for all prey.

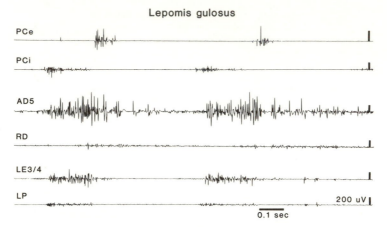

A

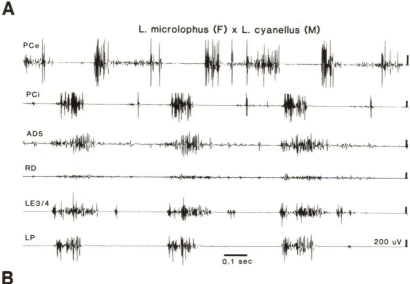

B

FIG. 1-6 Representative electromyographic recordings from pharyngeal muscles in **(A)** *Lepomis gulosus* and **(B)** a hybrid between *L. microlophus* and *L. cyanellus*. These data were obtained during swallowing of an earthworm (5 to 7 cm in length). All muscles in each panel were recorded simultaneously.

Discussion

Figure 1-8 summarizes the distribution of characters derived from the motor patterns in the pharyngeal muscles (black bars), morphological specializations in the pharynx (stippled bars), behavioral specializations (bars with vertical hatching), and ecological specializations (cross-hatched bars). The branching diagram depicting the relationships among the sunfish species is derived from other characters and is not fully resolvable with current information (note the four trichotomies). On the basis of this branching diagram, *Lepomis gibbosus* and *L. micro-*

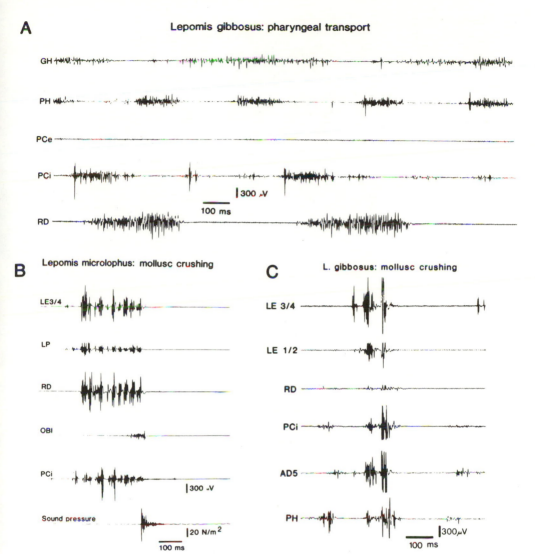

FIG. 1-7 Electromyographic recordings of pharyngeal muscles in *L. gibbosus* and *L. microlophus* during feeding on a worm (**A**) and on snails (**B** and **C**). Note the difference between the motor patterns in *L. gibbosus* in **A** and **C**. Snail-crushing patterns exhibit extensive overlap between pharyngeal muscles, in contrast to the rhythmic alternating pattern used for fish and worms. *L. microlophus*, however, uses the crushing pattern for *all* prey types (from Lauder, 1983a).

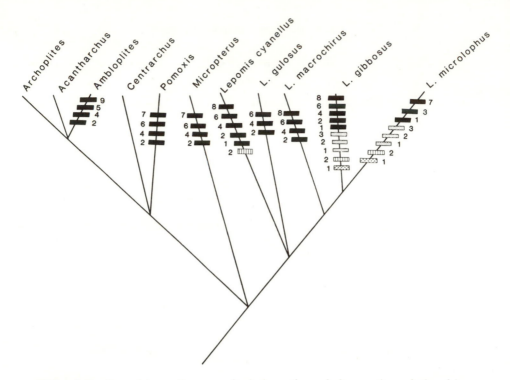

FIG. 1-8 Branching diagram depicting the phylogenetic relationships between selected centrarchid species (from Lauder, in preparation). Characters on the diagram are those resulting from the study of feeding behavior, morphology, and physiology, and were not used to generate the initial branching diagram. Bars are coded to indicate different types of characters. *Ecological features* (cross-hatched bars): 1, snails form a prominent component of the diet. *Behavioral features* (vertically hatched bars): 2, crushing behavior during feeding on snails. *Morphological features* (stippled bars): 1, hypertrophy of the pharyngohyoideus muscle; 2, hypertrophy of the levator posterior muscle; 3, expanded tooth areas on the upper and lower pharyngeal jaws. *Functional features* (black bars): 1, crushing motor pattern with extensive coactivation of pharyngeal muscles; 2, the retractor dorsalis and levator internus muscles alternate in activity during pharyngeal transport; 3, loss of rhythmic alternating motor pattern; 4, pharyngocleithralis internus has one burst of activity overlapping activity in the retractor dorsalis muscle; 5, pharyngocleithralis externus muscle has only one burst of activity, between adjacent bursts of the retractor dorsalis; 6, pharyngocleithralis externus muscle has two or three bursts, with extensive overlap in activity with the retractor dorsalis muscle; 7, pharyngohyoideus muscle has a double burst pattern with each burst overlapping activity in the retractor dorsalis; 8, triple burst pattern in the pharyngohyoideus muscle; 9, pharyngohyoideus muscle displays one burst of activity per swallowing cycle, alternating activity with the retractor dorsalis.

lophus are hypothesized to be sister groups (each other's closest relatives), and the other three *Lepomis* species on the diagram are proposed to be outgroups to this pair. Again, note that the numbered bars in Figure 1-8 are not used to support the branching diagram; they represent the distribution of characters derived from the case study on a previously corroborated phylogeny.

From this figure, we can consider that functional (black bar) characters 2, 4, and 6 are primitive for the clade, as they occur in all species except *L. microlophus* (*Ambloplites* lacks character 6). Thus, the phylogenetic distribution of these aspects of the motor pattern indicates that they represent a primitive component of the feeding behavior in this clade *and* that these characters are homologous in all taxa in which they are found. Other aspects of the motor pattern are specialized in some species. For example, *Ambloplites* has two unique features (numbers 9 and 5, black bars) that are not found in any other species. Functional character 7, occurring in *Pomoxis*, *Micropterus*, and *L. microlophus*, is, by itself, indicative of shared ancestry among these three taxa. However, this interpretation conflicts with the corroborated branching diagram and thus feature 7 must be convergent in at least one of the three taxa that possess it. The major specialization of motor patterns occurs in *L. microlophus* (character 3, black bar). This species has lost the rhythmic alternating motor pattern and uses the crushing pattern for all prey. The distribution of functional characters that relate to the rhythmic swallowing pattern indicates that possession of this pattern is primitive for the clade and homologous within it.

Within this clade, only two species have a prominent molluscan component to their diet, and the distribution of this ecological feature on the cladogram (character 1, cross-hatched bar) is congruent with the sister-species relationship of the pumpkinseed and redear sunfish. Also congruent with this branching pattern is the distribution of the three morphological specializations (characters 1, 2, 3, stippled bars). The most parsimonious interpretation of this congruent pattern is that a series of morphological, ecological, and behavioral novelties were acquired after the speciation event that gave rise to *L. macrochirus* and before or at the speciation event splitting the redear and pumpkinseed sunfish.

The final feature of interest is the crushing motor pattern in the pharyngeal muscles (character 2, vertically hatched bars). This motor pattern occurs in three species: *Lepomis microlophus*, *L. gibbosus*, and *L. cyanellus*. This motor pattern is homologous in the two snail-crushing species because it is congruent with the branching diagram, but it appears to be convergent in the green sunfish, *L. cyanellus*. Further information on the phylogenetic position of other *Lepomis* species and a resolution of the unresolved areas of the phylogeny are necessary to demonstrate that the green sunfish convergently possesses this character. However, it is clear from the distribution of this character in Figure 1-8 that the homology of this motor pattern within the clade is in doubt and that resolution of the issue depends on further phylogenetic analysis.

CONCLUSIONS

One major conclusion from this case study is that the distribution of functional, morphological, and behavioral features of a character complex are not necessarily congruent with each other. The implication is that each type of character is subject to convergence, and even complex motor patterns may be independently derived. From this example, it would appear that the neural criterion has little utility as an arbiter of behavioral homology. Within the sunfishes, morphological design in the pharyngeal region has remained conservative with no major alterations in pattern, and yet motor patterns controlling the pharyngeal muscles have changed greatly. Homologous muscles and other morphological elements have not retained their primitive pattern of activation and movement. The redear sunfish has completely lost the primitive motor pattern of rhythmic alternating activity, and although this species retains the primitive musculoskeletal design of other species in the Centrarchidae, a new motor pattern occurs that is unique to this species. No other teleost fish studied to date possesses such a novel motor pattern that is used for all prey. In this case, the peripheral morphology associated with the behavior appears to have remained relatively conservative, while the neural control has been modified considerably.

A second important point to emerge from this case study is that in the absence of functional information, a different interpretation might have been placed on the patterns. Yet, data on motor patterns are precisely what is lacking in so many comparative ethological investigations. In this case study, the major specializations and convergences within the clade were functional, and I think it highly unlikely that it would have been possible to predict a priori the distribution of motor programs within the clade based on morphology and behavior of the species. The data emphasize the extent to which transformations of motor patterns are involved in the evolution of behavioral specialization, and that "similar" behaviors may be mediated by very different motor programs.

In this chapter I have attempted to provide an analysis of a common (but not the exclusive) approach of ethologists to the analysis of homology. I have emphasized a priori avenues of ethological investigation because of the importance that these are usually accorded in general discussions and in research programs. I believe that arguments for homology in the absence of a direct link to a phylogenetic hypothesis are weak, and that the strongest test of homology is congruence with other characters. The phylogenetic approach has the added merit of providing a method of testing the assumptions about relative constancy of neural and morphological systems assumed by a priori arguments. An important future task of comparative ethologists must be to acquire more examples of phylogenetic patterns of behavior in which the distribution of both morphological and neural features related to the behaviors is included. Only by the examination of many such case studies will general patterns in the evolution of behavior emerge.

ACKNOWLEDGMENTS

Support for this research was provided by an A. W. Mellon Foundation fellowship, the Louis Block Fund (University of Chicago), and NSF DEB 81-15048 and NSF PCM 81-21649. Figures 1-4 and 1-5 were drawn by Clara Richardson. Many people provided assistance with the work described in this chapter or provided useful comments on the manuscript: S. Barghusen, P. Wainwright, W. Bemis, C. Smither, M. O'Shea, R. E. Lombard, P. S. Ulinski, and R. Eaton. I particularly thank D. Philipp of the Illinois Natural History Survey for his assistance in obtaining research animals, and J. Hopson, H. Greene, H. Barghusen, and C. Patterson for their critical comments on the manuscript.

LITERATURE CITED

Alcock, J. 1975. Animal Behavior. Sinauer Associates, Inc., Sunderland, Mass.

Arnold, S. J., and M. J. Wade. 1984a. On the measurement of natural and sexual selection: Theory. *Evolution* 38:709–719.

Arnold, S. J., and M. J. Wade. 1984b. On the measurement of natural and sexual selection: Applications. *Evolution* 38:720–734.

Aronson, L. R. 1981. Evolution of telencephalic function in lower vertebrates. In: Laming, P. R. (ed), Brain Mechanisms of Behaviour in Lower Vertebrates. Cambridge University Press, Cambridge, pp. 33–58.

Atz, J. W. 1970. The application of the idea of homology to animal behavior In: Aronson, L. R., E. Tobach, D. S. Lehrman and J. S. Rosenblatt (eds), Development and Evolution of Behavior: Essays in Honor of T. C. Schneirla. W. H. Freeman, San Francisco, pp. 53–74.

Ayers, J. L., and F. Clarac. 1978. Neuromuscular strategies underlying different behavioral acts in a multifunctional crustacean leg joint. *Journal of Comparative Physiology* 128:81–94.

Ayers, J. L., and W. J. Davis. 1977. Neuronal control of locomotion in the lobster, *Homarus americanus*. I. Motor programs for forward and backward walking. *Journal of Comparative Physiology* 115:1–27.

Baerends, G. P. 1958. Comparative methods and the concept of homology in the study of behavior. *Archives Neerlandaises de Zoolgie Supplement* 13:401– 417.

Barlow, G. W. 1977. Modal action patterns. In: Sebeok, T. A. (ed), How Animals Communicate. Indiana University Press, Bloomington, pp. 98–134.

Bateson, P. P. G., and R. A. Hinde. 1976. Growing Points in Ethology. Cambridge University Press, Cambridge.

Bertram, B. C. R. 1976. Kin selection in lions and in evolution. In: Bateson, P. P. G., and R. A. Hinde (eds), Growing Points in Ethology. Cambridge University Press, Cambridge. pp. 281–301.

Bock, W. and G., Von Wahlert. 1965. Adaptation and the form-function complex. *Evolution* 19:269–299.

Brown, J. 1975. The Evolution of Behavior. W. W. Norton, New York.

Bullock, T. H. 1961. The origins of patterned nervous discharge. *Behaviour* 17:48–59.

Bullock, T. H., D. A. Bodznick, and R. G. Northcutt. 1983. The phylogenetic distribution of electroreception: Evidence for convergent evolution of a primitive vertebrate sense modality. *Brain Research Reviews* 6:25–46.

Clutton-Brock, T. H., and P. H. Harvey. 1976. Evolutionary rules and primate societies. In: Bateson, P. P. G., and R. A. Hinde (eds), Growing Points in Ethology. Cambridge University Press, Cambridge, pp. 195–237.

Cracraft, J. 1981. The use of functional and adaptive criteria in phylogenetic systematics. *American Zoologist* 21:21–36.

Croll, R. P., and W. J. Davis. 1981. Motor program switching in *Pleurobranchia*. I. Behavioral and electromyographic study of ingestion and egestion in intact specimens. *Journal of Comparative Physiology* 145:277–287.

Delcomyn, F. 1980. Neural basis of rhythmic behavior in animals. *Science* 210:492–498.

Eberhard, W. G. 1982. Behavioral characters for the higher classification of orb-weaving spiders. *Evolution* 36:1067–1095.

Eibl-Eibesfeldt, I. 1970. Ethology, the Biology of Behavior. Holt, Rinehart, and Winston, New York.

Eldredge, N., and J. Cracraft. 1980. Phylogenetic Patterns and the Evolutionary Process. Columbia University Press, New York.

Fentress, J. C. (ed). 1976. Simpler Networks and Behavior. Sinauer Associates, Sunderland, Mass.

Gans, C., and W. J. Bock. 1965. The functional significance of muscle architecture—a theoretical analysis. *Ergebnisse der anatomie und entwicklungs-geschichte* 38:115–142.

Ghiselin, M. T. 1974. A radical solution to the species problem. *Systematic Zoology* 23:536–544.

Ghiselin, M. T. 1984. "Definition," "character," and other equivocal terms. *Systematic Zoology* 33:104–110.

Gosline, W. A. 1985. A possible relationship between aspects of dentition and feeding in centrarchid and anabantoid fishes. *Environmental Biology of Fishes.* 12:161–168.

Gould, J. 1982. Ethology. W. W. Norton and Company, New York.

Greene, H. W. 1977. Phylogeny, convergence, and snake behavior. Ph.D. thesis, University of Tennessee, Knoxville.

Greene, H. W., and G. M. Burghardt. 1978. Behavior and phylogeny: Constriction in ancient and modern snakes. *Science* 200:74–77.

Haas, O., and G. G. Simpson. 1946. Analysis of some phylogenetic terms, with attempts at redefinition. *Proceedings of the American Philosophical Society* 90:319–349.

Hailman, J. 1976a. Uses of the comparative study of behavior. In: Masterton, R. B., W. Hodos and H. Jerison (eds), Evolution, Brain, and Behavior: Persistent Problems. Lawrence Erlbaum Associates, Hillsdale, N.J., pp. 13–22.

Hailman, J. 1976b. Homology: Logic, information and efficiency. In: Masterton, R. B., W. Hodos, and H. Jerison (eds), Evolution, Brain, and Behavior: Persistent problems. Lawrence Erlbaum Associates, Hillsdale, N.J., pp. 181–198.

Heinroth, O. 1911. Beitrage zur Biologie, namentlich Ethologie und Psychologie derr Anatiden. Verhanalung des V Internationalen Ornithologen Kongresses, Berlin, 1910, pp. 589–702.

Hinde, R. A. 1966. Animal Behavior. McGraw-Hill, New York.

Hinde, R. A., and N. Tinbergen. 1958. The comparative study of species-specific behavior. In: Roe, A., and G. G. Simpson (eds), Behavior and Evolution. Yale University Press, New Haven, Conn., pp. 251–268.

Hodos, W. 1976. The concept of homology and the evolution of behavior. In: Masterton, R. B., W. Hodos, and H. Jerison (eds), Evolution, Brain, and Behavior: Persistent Problems. Lawrence Erlbaum Associates, Hillsdale, N.J., pp. 153–167.

Hoyle, G. 1976. Approaches to understanding the neurophysiological bases of behavior. In: Fentress, J. (ed), Simpler Networks and Behavior. Sinauer Associates, Sunderland, Mass., pp. 21–38.

Hull, D. 1978. A matter of individuality. *Philosophy of Science* 45:335–360.

Keast, A. 1978a. Trophic and spatial interrelationships in the fish species of an Ontario temperate lake. *Environmental Biology of Fishes* 3:7–31.

Keast, A. 1978b. Feeding interrelations between age-groups of pumpkinseed (*Lepomis gibbosus*) and comparisons with bluegill (*L. macrochirus*). *Journal of the Fisheries Research Board of Canada* 35:12–27.

Kessel, E. L. 1955. Mating activities of balloon flies. *Systematic Zoology* 4:97–104.

Lande, R., and S. J. Arnold. 1983. The measurement of selection on correlated characters. *Evolution* 37:1210–1226.

Lauder, G. 1981. Form and function: Structural analysis in evolutionary morphology. *Paleobiology* 7:430–442.

Lauder, G. 1982. Patterns of evolution in the feeding mechanism of actinopterygian fishes. *American Zoologist* 22:275–285.

Lauder, G. 1983a. Neuromuscular patterns and the origin of trophic specialization in fishes. *Science* 219:1235–1237.

Lauder, G. 1983b. Functional and morphological bases of trophic specialization in sunfishes (Teleostei: Centrarchidae). *Journal of Morphology* 178:1–21.

Lorenz, K. 1941. Vergleichende Bewegungsstidien an Anatinen. *Supplement, Journal of Ornithology* 89:194–294.

Lorenz, K. 1950. The comparative method in studying innate behavior patterns. *Symposium of the Society of Experimental Biology* 4:221–268.

Manning, A. 1967. An Introduction to Animal Behavior. E. Arnold, London.

Marler, P., and W. J. Hamilton. 1966. Mechanisms of Animal Behavior. John Wiley, New York.

Masterson, R. B., W. Hodos, and H. Jerison. 1976. Evolution, Brain, and Behavior: Persistent Problems. Lawrence Erlbaum Associates, Hillsdale, N.J.

Mayr, E. 1958. Behavior and systematics. In: Roe, A., and G. G. Simpson (eds), Behavior and Evolution. Yale University Press, New Haven, Conn., pp. 341–362.

Miller, R. J., and A. Jearld. 1982. Behavior and phylogeny of fishes of the genus *Colisa* and the family Belontiidae. *Behaviour* 83:155–185.

Miller, R. J., and H. W. Robison. 1974. Reproductive behavior and phylogeny in the genus *Trichogaster* (Pisces, Anabantoidei). *Zeitschrift fur Tierpsychologie* 34:484–499.

Nelson, G. J. 1970. Outline of a theory of comparative biology. *Systematic Zoology* 19:373–384.

Northcutt, R. G. 1981. Evolution of the telencephalon in nonmammals. *Annual Review of Neuroscience* 4:301–350.

Patterson, C. 1980. Cladistics. *Biologist* 27:234–240.

Patterson, C. 1982. Morphological characters and homology. In: Joysey, K. A., and A. E. Friday (eds), Problems of Phylogenetic Reconstruction. Academic Press, London.

Paul, D. H. 1981a. Homologies between body movements and muscular contractions in the locomotion of two decapods of different families. *Journal of Experimental Biology* 94:159–168.

Paul, D. H. 1981b. Homologies between neuromuscular systems serving different functions in two decapods of different families. *Journal of Experimental Biology* 94:169–187.

Pribram, K. 1958. Comparative neurology and the evolution of behavior. In: Roe, A. and G. G. Simpson (eds), *Behavior and Evolution*. Yale University Press, New Haven, Conn., pp. 140–164.

Roe, A., and G. G. Simpson (eds). 1958. Behavior and Evolution. Yale University Press, New Haven, Conn.

Savitz, J. 1981. Trophic diversity and food partitioning among fishes associated with aquatic macrophyte patches. *Transactions of the Illinois Academy of Science* 74:111–120.

Selverston, A. 1976. A model system for the study of rhythmic behaviors. In: Fentress J. (ed), Simpler Networks and Behavior. Sinauer Associates, Sunderland, Mass., pp. 82–98.

Simpson, G. G. 1958. Behavior and evolution. In: Roe, A., and G. G. Simpson (eds), Behavior and Evolution. Yale University Press, New Haven, Conn., pp. 507–535.

Tinbergen, N. 1959. Comparative studies of the behavior of gulls (Laridae): A progress report. *Behaviour* 15:1–70.

Ulinski, P. 1984. Design features of vertebrate sensory systems. *American Zoologist* 24:717–731.

Whitman, C. O. 1899. Animal Behavior. Biological Lectures of the Marine Biological Laboratory, Woods Hole, Mass.

Wiley, E. O. 1981. Phylogenetics: The Theory and Practice of Phylogenetic Systematics. John Wiley, New York.

2

Social and Unsocial Behavior in Dinosaurs

John H. Ostrom

*E*THOLOGY is a respected, albeit complex discipline that is securely based on direct observation and measurement. Of course, that does not mean that the conclusions reached by all observers are in full agreement, but at least direct observation of behavior is possible. Pity the poor paleo-ethologist who has no observational data of actual behavior—no record of time budgets, no record of time spent in foraging vs. resting, in hunting vs. courting, in guarding territory and clan—or just hanging around.

At first glance, speculating about behavior of any kind—in any kind of extinct animal—would seem to be an exercise in futility, a fool's errand. A pure ethologist no doubt would categorize speculations on dinosaur behavior as absurd—just fantasies. Nevertheless, at the request of our editors, I have been invited to expound on social (and other) behavior in dinosaurs. Let me say at the outset that most of what follows is based on work and ideas of others. These are assembled here for your consideration and entertainment.

The safest conclusion that I can come to is that dinosaur behavior must have been as diverse as the dinosaurs themselves, which came in many shapes and sizes. (Now having said that, I should stop writing.) Included were carnivores and herbivores, quadrupeds and bipeds, terrestrial kinds and others that are thought by some to have been at least amphibious, if not fully aquatic. Most were huge, as we all know, but some (although perhaps juveniles, or even hatchlings) were small—the smallest were perhaps the size of a robin. So we should anticipate a corresponding diversity of behavior.

For those who are not familiar with dinosaurian diversity, currently approximately 300 genera of dinosaurs have been named (not all of them wisely) and placed in one of two orders: the Saurischia and Ornithischia. Within these orders, traditional classifications list three suborders of Saurischians: the facultatively bipedal and largely, if not exclusively, herbivorous prosauropods; the huge quadrupedal and herbivorous sauropods like *Brontosaurus*; and the obligatory bipedal and carnivorous theropods. The Order Ornithischia consists exclusively of herbivorous kinds allocated to four or five subcategories: the facultative bipedal ornithopods (and sometimes separated near relatives, the pachycephalosaurs), the plated stegosaurs, the armored ankylosaurs, and the horned ceratopsians—all of which were obligate quadrupeds. Within this array, we can draw inferences about a few kinds of behavior for some and different kinds for others depending upon the quality of available evidence.

Exactly what is the nature of the evidence that pertains to dinosaurian behavior? To be brief, it falls into three categories: anatomic (the fossilized skeletal remains, usually very incomplete); taphonomic (the fossil associations and conditions of burial and preservation); and trace fossils (footprints and trackways). All of these are *indirect* evidence only. No behavior patterns or time budgets can be observed. From these indirect data we can only infer—and what you infer and what I infer from these may not be the same. But with rigorous analysis some inferences are testable with other evidence, and reasonable agreement can be achieved.

Abbreviations used here are as follows: AMNH, American Museum of Natural History; BPS, Bayerische Staatssammlung Paleontologie; and YPM, Yale Peabody Museum.

BEHAVIOR CATEGORIES

Behavior may be a solitary activity, or it may involve others of the same or different species. Both can be categorized into several distinct kinds, such as feeding, defense, movement (pursuit, escape, migration, etc.), courting and mating, nursery maintenance, and so on. Because evidence is not available for all the various possible activities of all the main dinosaurian varieties, the following exercise is organized by behavioral activities rather than by taxonomic groups.

Feeding Behavior

As Colbert (1958) observed, "there is a definite relationship between the morphology of an animal and its behavior. . . . Much of the behavior of animals is determined to a considerable degree by their physical adaptations." Consequently most of our inferences about the behavior of dinosaurs derive from their skeletal remains and inferred functional morphology. That is nowhere more evident than in their dentition, and that is the reason why we can say more about their feeding behavior than any other activity. As is evidenced by tooth morphology, there

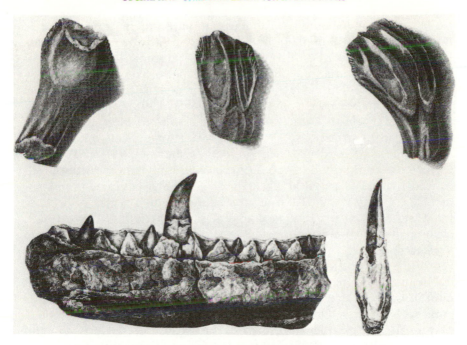

FIG. 2-1 Comparison of the serrated bladelike teeth of a flesh-eating thero-pod (bottom) in the lower jaw of *Megalosaurus* and the blunt-grinding tooth form of an herbivorous ornithopod *Iguanodon*. These illustrations are repro-duced from the earliest published illustrations of dinosaur remains (*Megalosau-rus* from Buckland, 1824; *Iguanodon* from Mantell, 1825).

were both flesh-eating and plant-eating dinosaurs. There is nothing new in this observation, but Figure 2-1 illustrates the obvious reasons for that conclusion.

The ornithischians all appear to have been herbivores, although recently discovered fragments that may be referable to the genus *Troo-don* might be an exception (D. Baird, 1980, personal communication). Nearly all ornithischian teeth were blunt without serrations and many show distinct occlusal grinding surfaces. Among the ornithischians, ste-gosaurs and ankylosaurs are enigmas regarding feeding habits or pref-erences. Both were bulky quadrupeds that carried their heads low, pre-sumably to browse on low, shrublike vegetation. Their jaws had broad, horny bills most likely used for plucking foliage, but the teeth behind were few in number and surprisingly small for such bulky animals. Whatever kind of vegetation they ate, it could not have been well chewed. Beyond this, we can deduce little about the feeding activities of these animals.

The bipedal ornithopods, on the other hand, were quite different. Their bipedal stance and progression could have increased their vertical foraging range, and may also have increased their running speeds. But it is their dentition that attracts attention. Early ornithopods of the Late Triassic and Earliest Jurassic, heterodontosaurs, featured surprising tooth differentiation: small, nipping, incisorlike front teeth, followed by prominent caninelike tusks, with dental batteries of special grinding teeth behind (Crompton and Charig, 1962). The larger ornithopods of Early to Late Jurassic times possessed large numbers of robust teeth that commonly display distinct wear facets indicative of some degree of food mastication (see Weishampel, 1984). The Late Cretaceous ornithopod varieties (those commonly known as duck-bills or hadrosaurs) featured highly specialized dental equipment and jaw mechanics (Figure 2-2). Their dentitions show a high degree of occlusal wear and efficient tooth replacement (Edmund, 1960; Ostrom, 1961), clear evidence of a sophisticated method of chewing food (see also Weishampel, 1984). This is surprising if one compares them with living reptiles. Some of the hadrosaurs reached large sizes, up to 5 m in bipedal height, and were well designed for browsing on high conifers abundant in Late Cretaceous forests. (Evidence of a possible preference for conifers is preserved in mummified remains of *Anatosaurus* now on display in the Senckenberg Museum of Frankfurt, West Germany, and reported by Krausel, 1922). Footprint evidence suggests that these animals may have browsed in groups (Lockley et al., 1983; Currie, 1983) among the singing pines.

By contrast, the related ceratopsians or horned dinosaurs were heavy quadrupeds with very large heads carried close to the ground. Here too, the jaws featured specialized dental batteries located behind cutting, parrotlike beaks. The teeth, although similar to those of the duck-bills, display vertical occlusal wear facets, which clearly indicate that the teeth were for shearing or slicing, rather than for grinding (Ostrom, 1964). We can only speculate about the preferred food, but it most probably was highly fibrous plant tissue, perhaps low-growing cycads or palms. That image of a cycad-browsing *Triceratops* may be enhanced by noting the "enlarged" size of the skull with its posterior bony extension or frill. In some ceratopsians, this frill is more than half the total skull length. Commonly the frill has been interpreted as a protective shield covering the vital neck region (Lull, 1908). Alternatively it has been interpreted as an expanded muscle-attachment site, allowing space for larger jaw muscles and adding power to the jaw-shearing mechanism (Haas, 1955; Ostrom, 1964). The frill may well have served both roles; in addition, Farlow and Dodson (1975) suggest that it had a display function.

The most demanding vegetarians among the dinosaurs must have been the giant sauropods. With weights ranging from perhaps 10 tons in the smallest species up to 60 tons or more in *Brachiosaurus* (Colbert, 1962), they must have been prodigious consumers, even if they were not endothermic (see Weaver, 1983). Paradoxically, they possessed no

FIG. 2-2 Dental battery of a hadrosaur (*Anatosaurus breviceps* YPM #1779) displaying the distinct occlusal surface (**A**), and the remarkable supply of replacement teeth (**B**) beneath the worn functional teeth.

obvious dental specializations or other adaptations that might have enhanced their feeding efficiency. Their elongated necks have traditionally (but not universally) been explained as an adaptation for elevating the head from a deep underwater position to permit breathing (for a debate on this, see Kermack, 1951; Colbert, 1952). A more realistic interpretation is that their long necks permitted browsing on high foliage (Bakker, 1968). That would seem to be an appropriate adaptation in the Jurassic-Early Cretaceous world, where nearly all other herbivores were low-level feeders (stegosaurs, ankylosaurs, and most ornithopods).

The carnivorous dinosaurs or theropods have formally been categorized (I think incorrectly) as large or small animals in two separate infra-orders: Carnosauria and Coelurosauria. That they all fed on flesh is evident from their teeth, which are slightly recurved, laterally compressed, and bladelike with serrated edges. With very few exceptions, there is no hard evidence as to whether any particular species was a predator or a scavenger. Likewise, for most kinds there is no evidence about hunting strategy or killing techniques.

Farlow (1976) speculated about the diet and foraging behavior of theropods, relying on analogies with Recent predators (crocodilians, the Komodo monitor, and several other lizards, mammals, and birds). The actual fossil evidence is sparse. Taphonomic evidence suggests that some theropods may have foraged in groups. For example, the famed *Coelophysis* Quarry at Ghost Ranch, New Mexico, contained numerous skeletons of the small carnivore *Coelophysis*, but very little else. The few other species recovered there (phytosaur fragments and three small thecodonts) were also carnivorous types. By contrast, the Cleveland-Lloyd Quarry of Utah has produced vast numbers of bones of *Allosaurus* of all size classes associated with several kinds of herbivores. The *Allosaurus* remains greatly outnumbered bones of the herbivorous kinds (*Camptosaurus* and *Stegosaurus*). Stokes (1961) has explained the disproportionate abundance of carnivore remains in the Cleveland-Lloyd Quarry as a Rancho-La-Brea type "predator trap" where *Allosaurus* was attracted in numbers to feed on mired-down dying or dead herbivores. The *Coelophysis* Quarry, however, is not so easily explained. There, only carnivores are preserved and they are almost exclusively the remains of *Coelophysis* at a ratio of about 30 to 1. The remains of *Coelophysis* include both young and adults and are preserved as articulated partial or complete skeletons. That suggests the expiration of a clan, perhaps at a drying-up water hole, but does not preclude their demise by flood at a clan scavenging feast. The taphonomic evidence is inconclusive.

While these two sites may appear ambiguous, there is persuasive evidence at another site that at least one kind of theropod, *Deinonychus*, probably hunted in packs. At the Yale *Deinonychus* Quarry in Montana, remains of at least four individuals (and perhaps that many more) of

Deinonychus were recovered associated with fragments of a single much larger herbivore, *Tenontosaurus* (Ostrom, 1969). The unabraded condition of the delicate *Deinonychus* remains, although often disarticulated but closely associated, argues that these remains were preserved at or very close to the site of death. A tempting conclusion that I accept is that these several 70 kg predators were killed during an attack on the much larger prey animal (ca. 600–700 kg), *Tenontosaurus*. Isolated, presumably shed, teeth of *Deinonychus* have been found associated with a number of other *Tenontosaurus* skeletons (Ostrom, 1970), suggesting that this particular ornithopod was favorite prey for the much smaller *Deinonychus*. The multiple remains at the Yale site and the pronounced size disparity between the two animals strongly indicate pack hunting by *Deinonychus*.

In contrast to group predation, there are at least two theropod specimens that could be interpreted as evidence of solitary predation: *Compsognathus* and *Coelophysis*. The classic specimen of chicken-size *Compsognathus* (BSP #A.S.I. 563) in the Bavarian State Collections in Munich clearly reveals the skeleton of a much smaller animal within its rib cage (Figure 2-3). Amazingly, this partial but still articulated skeleton is identifiable as that of an apparently cursorial lizard, *Bavarisaurus* (Ostrom, 1978). Although this lizard may have been flushed by group foraging, it obviously was caught and consumed by just one predator. (It appears to have caused fatal indigestion.) The evidence in *Coelophysis*, on the other hand, is not so clear. Within the body cavity of one adult skeleton (AMNH #7224) is a mass of disarticulated small bones that appear to be those of a very young *Coelophysis*, although that is not certain. It is not even clear whether these belong to one individual or include parts of several. This could be a case of group cannibalism, or perhaps scavenging, with these consumed incomplete remains evidence of more than one consumer.

The only convincing evidence of solitary predation among dinosaurs is footprint evidence reported from Texas (Bird, 1941), Queensland, Australia (Thulborn and Wade, 1979, 1984), and possibly Colorado (Lockley et al., 1983; Prince, 1983). In these reports, the authors note the occurrence of one or two trackways of large theropods paralleling trackways of various herbivorous dinosaurs. There is no way to establish whether these different trackways were made at the same time, but the fact that the predator trackway, in some instances, parallels a herbivore's trackway strongly suggests that one was stalking the other.

Concerning the killing tactics of the various theropods, there is very little evidence. But the bizarre anatomy of *Deinonychus* (Figure 2-4) provides us with remarkable clues about a peculiarly aggressive predator. Hunting in packs, as noted earlier, these animals apparently grasped the prey with their clawed forelimbs and slashed at vulnerable regions with a pair of large and sharp sicklelike claws on their hind feet. This hypothesized tactic (Ostrom, 1969) was confirmed by a remarkable dis-

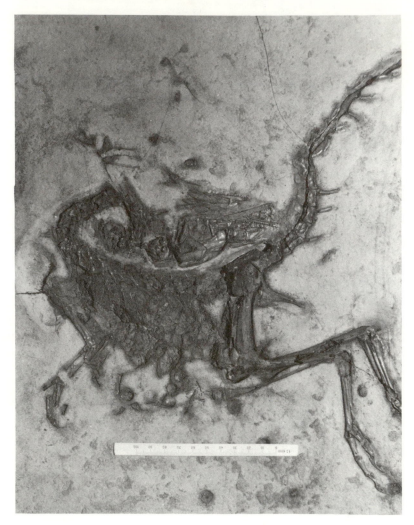

FIG. 2-3 Skeleton of the chicken-size theropod *Compsognathus* containing the articulated skeleton of a lizard *Bavarisaurus*. The proportions of the consumed lizard skeleton compare most closely with modern fast-running, ground-dwelling lizards like *Cnemidophorus*.

covery (as yet unreported in the scientific literature) in Mongolia of a near relative of *Deinonychus*—*Velociraptor*. It was preserved in fatal combat with its intended prey *(Protoceratops)*, the lethal pedal claw imbedded in the mid-section and the hands grasping the head of *Protoceratops*. To my knowledge, this is the only direct evidence available that clearly documents the method of kill by any theropod, and thus the only certifiable evidence of theropod predation, as opposed to scavenging.

FIG. 2-4 Reconstruction of the pack-hunting theropod *Deinonychus*. Notice the long claw-bearing hands and arms and the large sicklelike claw on each foot.

Speculations abound concerning the feeding habits of the giant theropods like *Tyrannosaurus* and *Tarbosaurus*. Their sheer size alone, together with their minuscule and seemingly useless forelimbs, suggests a scavenging mode, but we do not know. Yet it is difficult to visualize a 5-ton *Tyrannosaurus* succeeding in pursuit of any prey.

Mating Behavior

Not surprisingly, there is no evidence at all for the nature of dinosaur mating behavior. Clutches of eggs have been found of several varieties, notably the Mongolian ceratopsian *Protoceratops* (see Brown and Schlaikjer, 1940), the French sauropod *Hypselosaurus* (Matheron, 1869), and most recently the duck-bill *Maiasaura* from Montana (Horner and Makela, 1979; Horner, 1982) plus several other unidentifiable kinds (Horner, 1984). Horner's discoveries are among the most important and exciting dinosaur finds in decades. Not only has he recovered several different kinds of eggs in clutches; he has also found multiple nests of what appear to be the same kind in a single horizon, suggesting "colonial nesting." The same kind has also been found in different horizons. Horner interprets the latter as evidence of "site fidelity"— the gravid females returning to the same nesting site year after year (Horner, 1982, 1984). Even more important is Horner's discovery of nests of very young duck-bills apparently huddled together like bird hatchlings in a nest. But these young are too large to be very recent hatchlings. Moreover, their teeth show signs of wear. The question that cannot be answered is whether these young foraged for food on their own and returned to the shelter of their nest, or whether there is a suggestion here of parental care with the parents bringing food to the nestlings. Horner's discovery of this multi-species dinosaur nesting ground in Montana gives us a new window on dinosaurian biology.

The duck-bill dinosaurs have prompted the most speculation about dinosaur courting and mating behavior, chiefly because of their peculiar nasal apparatus and the variety of cranial crests in some (Figure 2-5). Hopson (1975) presented a convincing hypothesis (first suggested by Wiman, 1931) that the cranial crests (both solid and hollow varieties) of certain hadrosaurs were visual cues and that the hollow crests containing loops of the nasal passages were vocal resonating structures (see also Weishampel, 1981), all of which presumably promoted successful intraspecies identification and thus served as a premating, genetic-isolating mechanism. Hopson argued further that the large depressions enclosing the external nares in the crestless hadrosaurs housed "inflatable" diverticulae of the nasal passages that similarly served as display organs. Can't you see these creatures blowing their noses at each other

FIG. 2-5 The skull of *Corythosaurus casuarius* (AMNH #5240) with its prominent nasal crest. Notice the dental batteries.

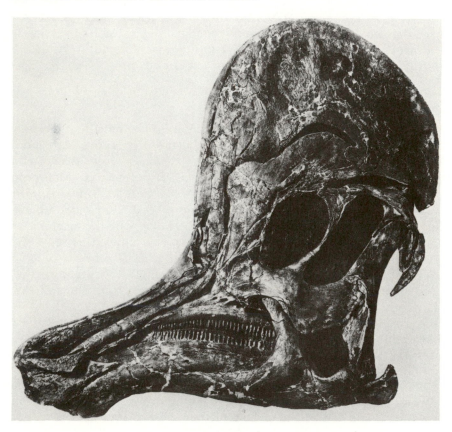

FIG. 2-6 The skull of *Triceratops brevicornus* (BS Munich, formerly YPM #1834) illustrating the frill extension and the facial horns. Notice also the dental batteries.

in search of a mate? Why not? Hadrosaurs had well-developed eyes (Ostrom, 1961) and ears (Colbert and Ostrom, 1958). Enhanced olfaction (Ostrom, 1962), perhaps provided by the expanded nasal tracts, may also have played a role in hadrosaurian species recognition (or approaching predators), but Weishampel (1981) has reinforced the resonating hypothesis. With a careful theoretical acoustical analysis, he argued on physical grounds that vocalization in the hollow-crested duck-bills was for parent-offspring communication rather than mate signaling. If one, why not both?

Turning to the horned dinosaurs, Farlow and Dodson (1975) pondered the variety of frill shapes and sizes together with variations in the nasal and brow horns (Figure 2-6). They concluded that these differences in cranial morphology reflect differences in intraspecific agonistic and courtship behavior, somewhat analogous to behavior in modern horned ungulates and some horned lizards. The earliest ceratopsians featured short frills and only a single nasal horn, or none at all. Farlow and Dodson suggested that the frill served as a "visual dominance rank symbol" and the nasal horn was used in intra-specific combat, with the snout and horn being swung against the flanks of the competitor. According to these authors, later ceratopsians had larger frills that enhanced the display function, and the more complex arrays of multiple facial horns were used in frontal combat with adversaries of their own kind. I like that scenario of a pair of rutting *Triceratops* bulls squaring off against each other to win potential mates or dominance of the group. It makes much more sense than the usual explanation of a threatened *Triceratops* fending off a hungry *Tyrannosaurus*.

Rutting behavior seems to have been true of another group of dinosaurs, the bipedal pachycephalosaurs, commonly known as "dome heads." Members of this group of ornithopods are characterized by massively thickened bony skull caps; Maryanska and Osmolska (1974) consider these to represent a separate sub-order quite distinct from the Ornithopoda. Galton (1970) has argued persuasively that this thickened bony dome was used in intraspecific contests in frontal head-butting analagous to that of American bighorn sheep. Presumably, this activity was by rutting males in competition for mates or to establish dominance over the herd.

Molnar (1977) has given an interesting review of the analogies in modern ungulate mammals and the various ornithischian dinosaurs to test some of the above inferences about structure and inferred behavior.

Defensive Behavior

Defensive strategy among dinosaurs undoubtedly was as varied as were the animals themselves. While the use of some structures seems obvious, sometimes the evidence is ambiguous. For instance, were the nasal and brow horns of the horned dinosaurs for active defense against predators, as frequently claimed, or were they for intraspecific sparring to establish dominance within the herd, as Farlow and Dodson (1975) have maintained? Analogy with living African antelopes suggests both. Among the armored dinosaurs, defense seems to have been chiefly passive, the animals being well shielded beneath thick bony scutes and spikes. Yet some, like *Ankylosaurus*, carried spikes or mace-like clubs at the end of the tail, suggesting a more active mode of defense. Probably even those without tail weapons thrashed about with the tail in defensive action.

The large, erect bony plates along the back of *Stegosaurus* have long been interpreted as defensive structures that made the animal appear larger in profile. Recent examination by Farlow et al. (1977) of the internal structure of these plates, combined with experimental studies, indicate that these bony structures probably served for thermal regulation rather than for protection, although Hopson (1977) suggested that they might have served both functions. They were highly vascularized and probably were heavily perfused with blood, probably to dissipate excess body heat. But *Stegosaurus* was also armed with large bony spikes at the end of the tail, which suggest an aggressive or active mode of defense.

SAUROPOD HERD at
DAVENPORT RANCH
TEXAS

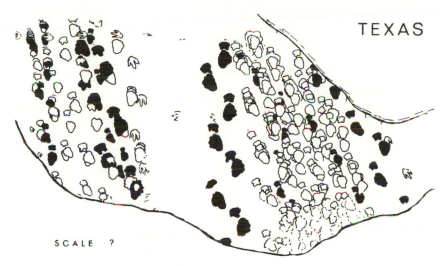

SCALE ?

From R.T. Bird, 1944

FIG. 2-7 Map of Bird's (1944) sauropod herd trackway field at Davenport Ranch, Texas. Notice the mix of large and small footprints, and the fact that they are all headed in approximately parallel traverses. Although Bakker (1968) claimed that the small footprints occur only in the center of the "herd" and the large footprints only on the periphery, that is not entirely evident even in the right-hand cluster where large and small animals occurred together near the center. In the left-hand cluster the evidence of herd structure is even more ambiguous.

Despite these apparently active defense adaptations, the predominant defensive behavior must have been by fleeing. Or perhaps safety in numbers was the dominant defensive strategy. Bird (1944) reported and illustrated a most informative series of parallel dinosaur trackways preserved in Early Cretaceous strata in Bandera County, Texas, which suggests group movement (Figure 2-7). Except for von Huene's (1928) suggestion of migrating behavior in the prosauropod *Plateosaurus*, this is the first documentation known to me of group activity (social or otherwise) that was founded on substantial evidence. Despite these notices, the idea of group activity in dinosaurs has received little published attention until recently, possibly because few believed that any evidence could document it. In 1972, the idea came to life again, resurrected by this author (Ostrom, 1972) in a description of a long-known site in Massachusetts, where several dozen trackways are preserved. I will return to that evidence later. It is obvious in living animals that flight from danger is the most common form of defense. To the best of my knowledge, there is only one clear paleontological example of this in dinosaurs—a footprint site in Queensland, Australia, reported by Thulborn and Wade (1979, 1984). They describe a series of trackways recording a "stampede" of more than 150 bipedal dinosaurs, identified as both ornithopods and coelurosaurs, that ran at speeds of up to 16 km per hour (estimated using Alexander's 1976 theorem). Associated with the stampede tracks is the trackway of a much larger theropod, an animal perhaps the size of *Tyrannosaurus*. Thulborn and Wade suggest that it was the presence of this large predator that triggered the flight of so many smaller animals, whose trackways are all closely parallel.

The 16 km per hour estimated speed of the fleeing Australian dinosaurs is not particularly impressive, but perhaps that was all that was required to avoid the grasp of the much larger and less fleet theropod. More importantly, that is significantly faster than the velocities estimated by Alexander (1976), Tucker and Burchette (1977), Kool (1981), Mossman and Sarjeant (1982), and Currie (1983) using Alexander's formula on trackway data at various sites, all of which indicate slow walking speeds of usually less than 10 km per hour. Such evidence of slow speeds is not surprising since animals walk much more often than they run, and the probability of preservation of the trackway of a running animal is far less than that recording a casual stroll. Coombs (1978), on anatomical evidence, and Thulborn (1982, 1984), on anatomical and trackway evidence, have theorized about cursorial speeds and gaits in a variety of dinosaurs and conclude that maximum running speeds in dinosaurs ranged from a low velocity of 6 to 7 km per hour in ankylosaurs and stegosaurs to a top speed of 43 km per hour in the ornithopod *Dryosaurus* and 56 km per hour in the coelurosaur *Gallimimus*. A recent report by Farlow (1981) of trackways at a site in Kimble County, Texas, appears to substantiate their conclusions. He documents trackways of three medium-size theropods that indicate velocities of nearly 30, 40, and 43 km per hour! Here is good evidence that some dinosaurs, while not as fleet as a thoroughbred race horse, were, as we all suspected, capable of respectable speeds of pursuit or escape.

Group or Social Behavior

Although instances of multiple dinosaur remains have been reported from a number of sites (for example, the Cleveland-Lloyd Quarry of Utah, the Carnegie Quarry at Dinosaur National Monument, the Yale Quarry #1 at Garden Park in Colorado, and the famed *Brachiosaurus* Quarries in Tanzania), these contained a variety of dinosaur kinds and appear to be post-mortem accumulations. There are, however, several mass assemblages that are intriguing because they are mono-specific. There is the mass burial of more than three dozen skeletons of the ornithopod *Iguanodon* recovered from a coal mine near Bernissart, Belgium, in 1878 (Dollo, 1882) and also the several dozen or more individuals of the prosauropod *Plateosaurus* recovered at Trossingen, West Germany (von Huene, 1928). The *Coelophysis* Quarry of New Mexico may be another example, except that it is not quite mono-specific. The two European sites have long been tacitly accepted as evidence of group congregation in those two varieties of dinosaurs. Von Huene (1928) even ventured to explain the concentration of *Plateosaurus* skeletons at Trossingen as mass mortality of a herd during migration. Convincing as these several assemblages may seem, indicating that social congregation occurred in at least some dinosaur species, it is still possible that these concentrations might have resulted from factors *other* than social assembly.

However you choose to assess those mass mortalities, the best evidence in support of gregarious habits among dinosaurs is found in the fossil footprint record—trace fossil data that have often been maligned. Many occurrences of multiple dinosaur tracks have been reported from around the world, but, with the exception of Bird (1944), until recently no inferences had been drawn about possible gregarious behavior in dinosaurs from such footprint evidence. Bird reported and illustrated a remarkable site on Davenport Ranch in Bandera County, Texas, that revealed trackways of several dozen sauropods with nearly parallel orientation. The footprints record young as well as adult-sized animals. Bird remarked that "all were headed toward a common objective" and suggested "that they passed in a single herd"! That certainly appears to have been the case.

In 1968, Bakker went one step further. He noted that "these animals were not merely a disorganized mob of reptiles" (which Bird had neither stated nor implied) "but rather they were socially arranged in what appears to have been a true herd. The very largest footprints were made only at the periphery of the herd; the very smallest were made only in the center of the herd." While there may be some truth to Bakker's structured-herd interpretation of the Davenport Ranch evidence, that evidence is not as free of ambiguity as Bakker's statements assert (see Langston, 1974). Bird's map of the Davenport Ranch track field clearly shows that (see Figure 2-7).

Structured or not, the herding behavior of dinosaurs was first shown and recognized by Bird at Davenport Ranch and was substantiated by the remarkable record preserved at Holyoke, Massachusetts (Ostrom, 1972), which shows the traverses of 28 individuals, 20 of which are nearly parallel, trending in a generally westerly direction (Figure 2-8). All of the nearly parallel trackways appear to have been made by the same kind of bipedal animal, to which the footprint name *Eubrontes* has been applied. Of the eight other trackways that do not parallel the group, half appear to have been made by a different kind of dinosaur. The conclusion seems inescapable: here is clear evidence of a herd of one species of dinosaur strolling together across the Connecticut Valley landscape.

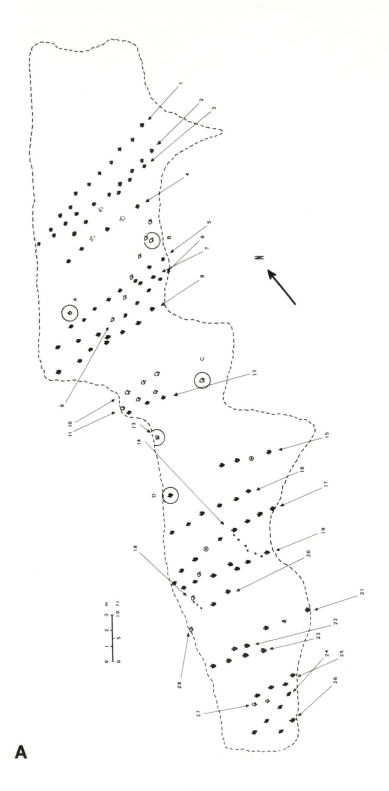

A

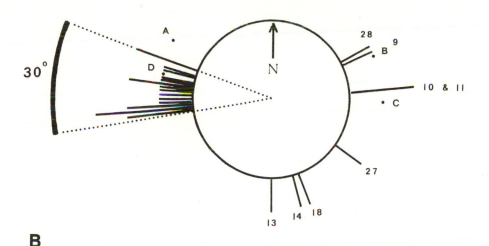

B

FIG. 2-8 (**A**) Map of the dinosaur trackways at the Mt. Tom site in Holyoke, Massachusetts. Arrows indicate the general trends and the numerical designations assigned to individual trackways. Solitary prints (encircled) that are believed to be continuations of trackways are designated by letters. Small open circles indicate former positions of removed or destroyed prints. (**B**) The diagram plots the trackway (solid lines) and solitary footprint (dots) bearings at the Mt. Tom site. Lengths of the solid lines are proportional to the number of trackways with that bearing (maximum of three). All of the trackways enclosed within the 30-degree arc appear to be of the same kind—*Eubrontes*. Of the other trackways, numbers 14, 18, 27, 28, and print A are of a distinctly different kind—*Grallator* and *Anchisauripus*. The evidence clearly points to group behavior in whatever animals made the *Eubrontes* trackways.

Since the Holyoke site was reported, several other similar records have been recognized, and previously reported sites have been reexamined. These include Albritton's (1942) site at Comanche, Texas; Calkin's (1933) site in Dorset, England; and Charig and Newman's (1962) locality in Swanage, England. More recently, Mossman and Sarjeant (1982), Currie (1983), Currie and Sarjeant (1979), Lockley et al. (1983), Lockley (1984), and Prince (1983) have reported comparable trackway evidence at sites in British Columbia and Colorado. These multiple records of near-parallel traverses by numerous individuals are the most convincing evidence available that several kinds of dinosaurs did in fact congregate and move in groups. But there are some question marks. For example, a few sites, particularly the F6 Ranch locality in Texas reported by Farlow (1981), record noticeably "symmetrical" traverses, with most trackways oriented in opposing directions with either Northwest or Southeast bearings. There is no evidence (Farlow, 1983, personal communication) that any barriers existed that confined those travelers to a "sidewalk" pathway, but that possibility cannot be ruled out. Yet the number of trackway sites around the world that show preferred trackway orientation is surprising. We can only speculate on these intriguing sites and what they portray about social behavior and how structured or unstructured dinosaurian community life may have been.

CONCLUSIONS

Of the several kinds of evidence available—anatomic, taphonomic, and trace fossils—the last evidence presents the least ambiguous picture of what we might identify as "social" behavior in dinosaurs. Footprint trails at numerous sites around the world point repeatedly (but not with absolute certainty) to group activity in several kinds of dinosaurs, ranging from Late Triassic to Late Cretaceous in age. These data are not easily dismissed—nor are they proof. Should we refer to the perpetrators as herds or as flocks?

LITERATURE CITED

Albritton, C. C., Jr. 1942. Dinosaur tracks near Comanche, Texas. *Field and Laboratory* 10:161–181.

Alexander, R. M. 1976. Estimates of speeds of dinosaurs. *Nature* 261:129– 130.

Bakker, R. T. 1968. The superiority of dinosaurs. *Discovery* (New Haven) 3(2):11–22.

Bird, R. T. 1941. A dinosaur walks into the museum. *Natural History* 47(2):74–81.

Bird, R. T. 1944. Did *Brontosaurus* ever walk on land? *Natural History* 52(2):60–67.

Brown, B., and E. M. Schlaikjer. 1940. The structure and relationships of *Protoceratops*. *Annals of New York Academy of Science* 40:133–266.

Buckland, W. 1824. Notice on the *Megalosaurus*. *Transactions of the Geological Society of London* 2(1):390–396.

Calkin, J. B. 1933. *Iguanodon* footprints in Dorset. *Discovery* (London) 14:13.

Charig, A. J., and B. H. Newman. 1962. Footprints in the Purbeck. *New Scientist* 14:234–235.

Colbert, E. H. 1952. Breathing habits of the sauropod dinosaurs. *Annual Magazine of Natural History* 5:708–710.

Colbert, E. H. 1958. Morphology and behavior. In: Roe, A. and G. G. Simpson (eds), *Behavior and Evolution*. Yale University Press, New Haven, Conn., pp 27–47.

Colbert, E. H. 1962. The weights of dinosaurs. *American Museum of Natural History Novitates* 2076:1–16.

Colbert, E. H., and J. H. Ostrom. 1958. Dinosaur stapes. *American Museum of Natural History Novitates* 1900:1–20.

Coombs, W. P. 1978. Theoretical aspects of cursorial adaptations in dinosaurs. *Quarterly Review of Biology* 53:393–418.

Crompton, A. W., and A. J. Charig. 1962. A new ornithischian from the Upper Triassic of South Africa. *Nature* 196:1074–1077.

Currie, P. J. 1983. Hadrosaur trackways from the Lower Cretaceous of Canada. *Acta Palaeontologica Polonica* 28:63–73.

Currie, P. J., and W. A. S. Sarjeant. 1979. Lower Cretaceous dinosaur footprints from the Peace River Canyon, British Columbia, Canada. *Palaeogeography, Palaeoclimatology, Palaeoecology* 28:103–115.

Dollo, L. 1882. Premiere note sur les dinosauriens de Bernissart. *Bulletin du Musee royal d'histoire naturelle de Belgique* 1:161–180.

Edmund, A. G. 1960. Tooth replacement phenomena in the lower vertebrates. *Contributions of the Royal Ontario Museum, Life Science Division* 52:1–190.

Farlow, J. O. 1976. Speculations about the diet and foraging behavior of large carnivorous dinosaurs. *American Midland Naturalist* 95:186–191.

Farlow, J. O. 1981. Estimates of dinosaur speeds from a new trackway site in Texas. *Nature* 194:747–748.

Farlow, J. O., and P. Dodson. 1975. The behavioral significance of frill and horn morphology in ceratopsian dinosaurs. *Evolution* 29:353–361.

Farlow, J. O., C. V. Thompson, and D. E. Rosner. 1977. Plates of the dinosaur *Stegosaurus*: forced convection heat loss fins? *Science* 192:1123–1125.

Galton, P. M. 1970. Pachycephalosaurids—dinosaurian battering rams. *Discovery* (New Haven) 6(1):23–32.

Haas, G. 1955. The jaw musculature in *Protoceratops* and in other ceratopsians. *American Museum of Natural History Novitates* 1729:1–24.

Hopson, J. A. 1975. The evolution of cranial display structures in hadrosaurian dinosaurs. *Paleobiology* 1:21–43.

Hopson, J. A. 1977. Relative brain size and behavior in archosaurian reptiles. *Annual Review of Ecology and Systematics* 8:429–448.

Horner, J. R. 1982. Evidence of colonial nesting and site fidelity among Ornithischian dinosaurs. *Nature* 297:675–676.

Horner, J. R. 1984. The nesting behavior of dinosaurs. *Scientific American* 241(4):130–137.

Horner, J. R., and R. Makela. 1979. Nest of juveniles provides evidence of family structure among dinosaurs. *Nature* 282:296–298.

Kermack, K. A. 1951. A note on the habits of the sauropods. *Annual Magazine of Natural History* 4:830–832.

Kool, R. 1981. The walking speed of dinosaurs from the Peace River Canyon, British Columbia, Canada. *Canadian Journal of Earth Science* 18:823–825.

Krausel, R. 1922. Die Nahrung von *Trachodon*. *Palaeontologische Zeitschrift* 4:80.

Langston, W., Jr. 1974. Nonmammalian Comanchian tetrapods. *Geoscience and Man* 8:77–102.

Lockley, M. G. 1984. Dinosaur tracking. *Science Teacher*. 1984(1):18–24.

Lockley, M. G., B. H. Young, and K. Carpenter. 1983. Hadrosaur locomotion and herding behavior: Evidence from the Mesa Verde Formation, Grande Mesa Coal Field, Colorado. *Mountain Geologist* 20:5–14.

Lull, R. S. 1908. The cranial musculature and the origin of the frill in the ceratopsian dinosaurs. *American Journal of Science* 25:387–399.

Mantell, G. 1825. Notice on the "*Iguanodon.*" *Philosophical Transactions of the Royal Society of London* 65:179–186.

Maryanska, T., and H. Osmolska. 1974. Pachycephalosauria, a new suborder of Ornithischian dinosaurs. *Paleontologia Polonica* 30:45–102.

Matheron, P. 1869. Note sur les reptiles fossiles des depots fluvio-lacustres cretaces du bassin a lignite de Fuveau. *Bulletin de la Societe Geologique de France* 26(2):781–795.

Molnar, R. E. 1977. Analogies in the evolution of combat and display structures in ornithopods and ungulates. *Evolutionary Theory* 3:165–190.

Mossman, D. J., and W. A. S. Sarjeant. 1982. The footprints of extinct animals. *Scientific American* 248:74–85.

Ostrom, J. H. 1961. Cranial morphology of the hadrosaurian dinosaurs of North America. *Bulletin American Museum of Natural History* 122:33–186.

Ostrom, J. H. 1962. The cranial crests of hadrosaurian dinosaurs. *Yale Peabody Museum of Natural History Postilla* 62:1–29.

Ostrom, J. H. 1964. A functional analysis of jaw mechanics in the dinosaur *Triceratops*. *Yale Peabody Museum of Natural History Postilla* 88:1–35.

Ostrom, J. H. 1969. Osteology of *Deinonychus antirrhopus*, an unusual theropod from the Lower Cretaceous of Montana. *Bulletin Yale Peabody Museum of Natural History* 30:1–165.

Ostrom, J. H. 1970. Stratigraphy and paleontology of the Cloverly Formation (Lower Cretaceous) of the Bighorn Basin area, Wyoming and Montana. *Bulletin Yale Peabody Museum of Natural History* 35:1–234.

Ostrom, J. H. 1972. Were some dinosaurs gregarious? *Palaeogeography, Palaeoclimatology, Palaeoecology* 11:287–301.

Ostrom, J. H. 1978. The osteology of *Compsognathus longipes* Wagner *Zitteliana* 4:73–118.

Prince, N. K. 1983. Late Jurassic dinosaur trackways from SE Colorado. *University of Colorado at Denver, Geology Department Magazine* 2:15–19

Stokes, W. L. 1961. Dinosaur Quarry near Cleveland, Utah. *Proceedings Utah Academy of Science* 38:132–133.

Thulborn, R. A. 1982. Speeds and gaits of dinosaurs. *Palaeogeography, Palaeoclimatology, Palaeoecology* 38:227–258.

Thulborn, R. A. 1984. Preferred gaits of bipedal dinosaurs. *Alcheringa* 8:243–252.

Thulborn, R. A., and M. Wade. 1979. Dinosaur stampede in the Cretaceous of Queensland. *Lethaia* 12:275–279.

Thulborn, R. A., and M. Wade. 1984. Dinosaur trackways in the Winton Formation (Mid-Cretaceous) of Queensland. *Memoirs of the Queensland Museum* 21-2:413–517.

Tucker, M. E., and T. P. Burchette. 1977. Triassic dinosaur footprints from South Wales: Their context and preservation. *Palaeogeography, Palaeoclimatology, Palaeoecology* 22:286–291.

Von Huene, F. 1928. Lebensbild der Saurischer-Vorkommens in obersten Keuper von Trossingen in Wurttemberg. *Palaeobiologie* 1:103–116.

Weaver, J. C. 1983. The improbable endotherm: The energetics of the sauropod dinosaur *Brachiosaurus*. *Paleobiology* 9:173–182.

Weishampel, D. B. 1981. Acoustic analyses of potential vocalization in lambeosaurine dinosaurs (Reptilia: Ornithischia). *Paleobiology* 7:252–261.

Weishampel, D. B. 1984. Evolution of jaw mechanisms in ornithopod dinosaurs. *Advances in Anatomy and Cell Biology* 87:1–110.

Wiman, C. 1931. *Parasaurolophus tubicen*, n. sp. aus der Kreide in New Mexico. *Nova Acta Regiae Societas Scientiarum Upsaliensis Series* 4 7(5):1–11.

3

Evolution of Behavior as Expressed in Marine Trace Fossils

Adolf Seilacher

*I*F we define behavior as the rules, or programs, underlying animal activities (i.e., if we exclude locomotion mechanisms and all growth programs, no matter how plastic and adaptive they may be), little seems to be left for a paleontologist to contribute to the theme of this volume. Direct evidence of ancient behavior is then restricted to patterns of larval attachment in sessile organisms, to bite marks or borings of predators on hard parts of other organisms (J. A. Kitchell, Chapter 4 herein) and to trace fossils left in soft sediments (J. A. Ostrom, Chapter 2 herein). It is the trace fossils, more exactly the burrows, of marine invertebrates that this chapter is concerned with.

Instead of primarily asking the traditional "whodunit" questions, we shall explore them as a record of highly organized behavioral programs that have evolved in certain sediment feeders of unknown affiliation. Although being still performed today, these activities are difficult to observe because they are mostly in the deep sea and their traces tend to be incomplete and difficult to see in sediment cores, while they show beautifully in the fossil record. Thus, in spite of the taxonomic uncertainties, trace fossils have, for a student of behavioral evolution, the double advantage that they are easily accessible and that they record changes of behavior patterns over hundreds of millions of years. In such a view it imports little that we can rarely refer these fossils to special kinds of animals, as long as genetically related groups can be singled out by diagnostic criteria. In large part this chapter reiterates what has been said in previous publications, but it supports these ideas with largely new illustrations.

Before we can go on to discuss particular case histories, a few basic principles must be introduced.

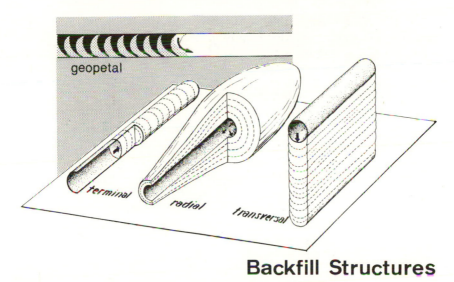

geopetal

terminal   radial   transversal

Backfill Structures

FIG. 3-1 The ways in which sediment feeders stuff the processed sediment back provide useful criteria for the classification of trace fossils, in which these structures are enhanced by preservational processes. In addition to the figured basic types, various combinations are found in nature. The geopetal effect (coarser-grained lamellae tend to slide down gravitationally more than the muddy ones shown in black) is important not only to tell tops and bottoms, but also to understand different preservational expressions of the same trace fossil.

Intrastratal origin. Surface traces as observed in modern tidal flats or in bottom photographs can be used as keys to the understanding of trace fossils, but only in an indirect way. Because of their low fossilization potential, invertebrate surface traces in the strict sense are virtually unknown in the fossil record. What we find of arthropod footprints, for instance, are usually undertracks impressed on an interface a few millimeters below the actual sediment surface (Goldring and Seilacher, 1971). Similarly, "worm trails" on sandstone beds can usually be shown (for instance by steeper contours than would be stable at the surface) to be burrows along the interface with under- or overlying mud.

Backfill structures. Sediment feeders do not simply wedge their way through the sediment, but rework it along a cross-section commonly exceeding their own and stuff the processed material back in an organized, laminated fashion. Of the basic types of backfill structures (Figure 3-1), many characteristic modifications are found in nature that can be used as "fingerprints" of the unknown burrowers. The appearance of the trace fossil may also be modified by the geopetal effect shown in Figure 3-1. It makes the originally crescentic sandy laminae slide down so that the upper half of the backfill becomes mainly muddy. As a result, the same burrow may look very different if preserved on the upper or the lower face of a sandy layer (see below). Such preservational modifications have led to the creation of many unnecessary names in the past.

Environmental control. Infaunal behavior is largely a response to environmental conditions. In the marine realm, we observe a significant trend to produce more systematic and complicated burrow patterns with increasing water depth. Since trace fossils (in contrast to body fossils) cannot be reworked, this trend is a useful tool to distinguish shallow marine and deep-sea sediments (Seilacher, 1967a). It is also because of this trend that the evolution of behavior patterns can best be studied in trace fossils from deep-sea sediments as represented by the flysch facies of folded mountain belts. Most of the trace fossils discussed here come from this facies.

OPTIMIZATION OF MEANDERING BEHAVIOR

Meandering is the most efficient method to cover, or scan, a given surface in a continuous way. Humans use it when ploughing fields, cutting lawns, and shading drawings after they have outgrown the scribbling stage. So do many animals. It is probably no coincidence that systematically scribbling traces are found only in rocks of early Paleozoic age (Seilacher, 1967b).

Regular meanders, in which subsequent lobes are guided by previous ones without overcrossing, are presently made in a variety of environments by a diversity of organisms and for very different purposes (Figure 3-2). In fossil meander trails and burrows, however, food extraction from the sediment is the only feasible purpose. The striking similarity of the meander patterns shown in Figure 3-3 is due to convergence, since their backfill structures are very different. Like some examples in Figure 3-2, they formally represent a combination of two patterns, a spiral followed by meanders. But such a descriptive approach is inadequate.

In order to understand these patterns in terms of behavior, it is useful to translate them into simple programs. To make a spiral in a two-dimensional space, for instance, only one order is needed:

(1) "Follow the previous trail at about the same distance d." This order implies the taboo against both overcrossing and leaving the particular bedding plane. For the production of meanders an additional order is required.

(2) "Turn back after having covered a distance l." This order must be stronger, because it must for a moment override the first order.

The validity of such an approach has been demonstrated by com-

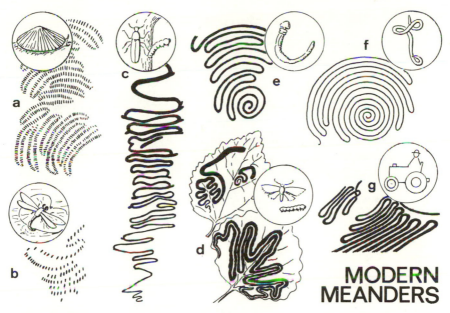

MODERN MEANDERS

FIG. 3-2 Modern examples of meandering behavior reflect a variety of biological purposes with the common denominator of continuously and efficiently covering a given surface. But not even the tracks left by sediment feeders on the deep-sea floor (**e**) are direct analogues of meandering trace fossils, because they are made at the surface rather than inside the sediment. (**a**) Radular feeding trace of a limpet. (**b**) Eggs laid by dragon fly on leaf of a waterlily. (**c**) Larval tunnel of wood beetle under bark. (**d**) Larval mines of moth in leaves. (**e**) Fecal string of enteropneust on deep-sea floor. *Nereites* (see Figure 3-5) might be the product of infaunal animals similar to this one. (**f**) Tunnel of annelid worm (*Paraonis*) in beach sand. (**g**) Track of tractor spraying a field (aerial picture) (Modified from Frey and Seilacher, 1980.)

1969) strikingly resemble natural patterns (Figure 3-4). They also show that (a) the starter spiral needs no particular programming, but is the outcome of a situation in which no straight stretch is as yet available for a reference, and that (b) new pattern qualities can be produced by simply changing the ratio of *d* to *l* in the same program.

Other computer experiments (Papentin and Röder, 1975) showed that the meander program itself can evolve from random behavior by simple Darwinian selection against trail overcrossing, although the assumptions used in the simulation may be oversimplified. By "punishing" poorly performing individuals with a reduced reproduction rate, Papentin and Röder got more and more guided patterns in successive generations (Figure 3-4). The outcome was the same (convergence) if the initial forms moved initially mainly straight ahead (computer species *Rectangulus rectus*, Figure 3-4) or made 90-degree turns in most steps (*Rectangulus vagus*). Papentin and Röder's computer evolution

FOSSIL MEANDER BURROWS

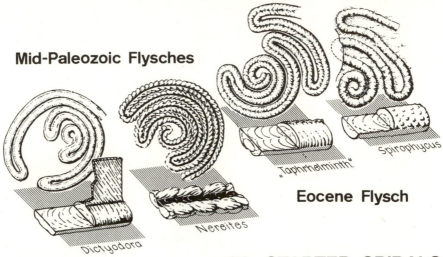

Mid-Paleozoic Flysches

Spirophycus

"Taphrhelminth"

Eocene Flysch

Nereites

Dictyodora

with STARTER SPIRALS

FIG. 3-3   Meandering trace fossils can be classified after their different back-fill structures (block diagram). Their geometrically similar patterns, with a starter spiral and subsequent guided meanders, reflect the convergent evolution of similar behavioral programs in sediment feeders of diverse affiliation in ancient deep-sea environments (flysch facies). Note that *"Taphrhelminthopsis"* is a preservational variant of *Scolicia* (see Figure 3-6) (Modified from Frey and Seilacher, 1980.)

also reflects the importance of chance effects: different lineages randomly resulted in either more spiral or more meandering patterns.

In matching this theoretical knowledge to actual phylogenetic lineages, however, we shall see that pattern optimization has worked not so much on the program itself as on its execution by improved abilities to turn sharply and to monitor the distance *d* from previous lobes of the burrow.

NEREITES *"LINEAGE"*

Nereites is the burrow of a worm-like sediment feeder that sorted out the coarse sediment particles with its front end and stowed them away into backfill lobes on all sides around the median tunnel. Anatomically, enteropneusts (Figure 3-2e) with their expandable protosoma are the favorite candidates for authorship. Because of the geopetal interfin-

COMPUTER SIMULATION

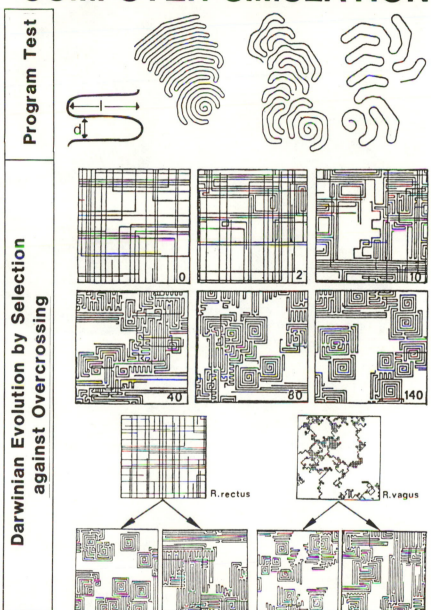

Program Test

Darwinian Evolution by Selection against Overcrossing

R.rectus

R.vagus

FIG. 3-4   Prescription of only the distance d and the length l of the turns, allowing for stochastic deviations, suffices to produce naturalistic meander patterns in the computer. Different ratios of d/l cause quantitative differences in the length of the starter spiral and in the way accidental marginal contouring of previous turning points changes the overall direction. (Top row from Raup and Seilacher, 1969. Bottom row from Papentin, 1975.)

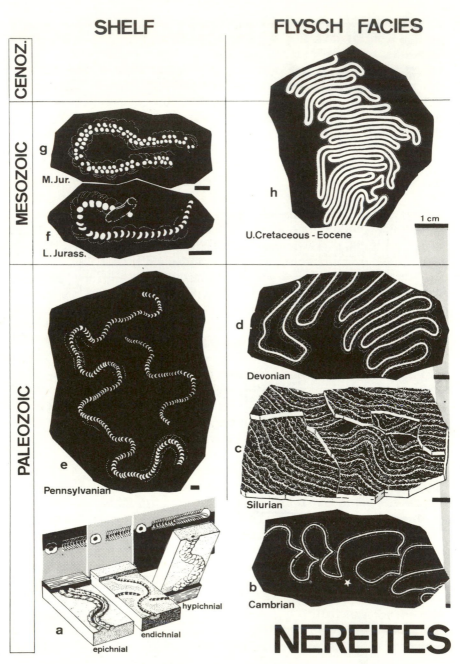

SHELF

FLYSCH FACIES

CENOZ.

MESOZOIC

g
M. Jur.

f
L. Jurass.

h
U. Cretaceous - Eocene

1 cm

PALEOZOIC

e
Pennsylvanian

d
Devonian

c
Silurian

b
Cambrian

a
epichnial
endichnial
hypichnial

NEREITES

FIG. 3-5 The ichnogenus *Nereites* is characterized by a complex backfill structure whose aspect changes in different modes of preservation (**a**). While meandering remains rather irregular in shallow marine (shelf) environments (**e–g**)
it becomes denser and more systematic in the flysch facies of deep-sea origin
(**b–d, h**). In the latter one also observes a gradual decrease in size. The kinks
in the early representative (**b**) reflect a physiological difficulty of continuously

gering of clay and sand laminae, the radial-terminal backfill structure of the lateral lobes can best be seen in the epichnial preservation on the top of sandy layers (*Phyllodocites* preservation, Figure 3-5a). At the same time the food-rich finer fraction was being passed through the gut and stuffed into the median tunnel behind the animal, either in the form of fecal packages or meniscoid segments (terminal backfill, Figure 3-1).

Small burrows associated with the Ediacara body fossils of the late Precambrian may be the earliest representation of *Nereites*, but they are too small to show the distinctive backfill structures. Even more tantalizing is the possibility that an associated form (Glaessner, 1969, Figure 5c, d; Fedonkin, 1981, Plate 15, Figure 3) is already a tightly guided meander version of *Nereites*, a type that by later times is found only in deep-sea environments. But again, the level of resolution is too low to be certain.

Because of its larger size, the next occurrence in Cambrian flysch of Argentina is more reliable (Figure 3-5b). Its conspicuous kinks reveal problems in the execution of the meander program. The turns brought the animal too far away from the previous lobe to monitor the distance *d*. So it curved back in the general direction until it hit the previous turn, but avoided overcrossing by sharply turning away and preparing for the next loop. In the absence of permanent monitoring, this "shock" had become such an essential element that it was released even if the animal failed to touch the previous lobe (asterisk in Figure 3-5b).

Later in the Phanerozoic, we find *Nereites* in shallow marine as well as in flysch sediments, but with different behavior. In the neritic facies, guided meanders are virtually absent, but we find characteristic modifications in the median string, in which the fecal packages are either meniscoid (endichnial *Scalarituba* preservation, Figure 3-5g) or laid down as round bodies in either a serial or a biserial, alternating fashion (*Neonereites*, Figure 3-5f, g).

In the flysch facies, in contrast, we observe the evolution of increasingly denser meanders without kinks. Somewhat out of the phase, the densest packing is reached already by a Silurian form, in which the ratio of l/d has become so high that not a single turn is contained in the published specimens (Chamberlain, 1978 Figure 3-5e). Whether the again lower l/d ratio in Devonian and Carboniferous flysches (Figure 3-5f) marks an evolutionary loss in efficiency or the extinction of the more specialized Silurian forms remains unclear.

---

sensing the distance to the next turn. Note that this kink was necessary to initiate the next turn even in the case that the animal failed to touch the previous trail (star) (**b, d, h** modified from Frey and Seilacher, 1980, Figure 14; **c** redrawn from Chamberlain, 1978; **e** after field photograph, Atoka Fm., Arkansas; **f** and **g** after Seilacher, 1960).

Another peak in meander density is reached by *Helminthoida* in Upper Cretaceous and Tertiary flysches (Figure 3-6h). This form, however, has become extremely miniaturized. It is also found not in sandy layers but in the chalky oozes formed by the newly evolved calcareous plankton. In such sediments, we usually see only the median string of fecal backfill, but rare specimens also show the oblique backfill of the marginal lobes that are in close contact with the ones of the previous loop.

SCOLICIA "LINEAGE"

The term *Scolicia* is used for a variety of burrows that have in common a depressed, elliptical cross-section and a terminal backfill of densely packed meniscoid lamellae. Such structures are presently produced by irregular echinoids (Spatangoidea) that plough through the sediment by moving particles from in front along the body surface and stuffing them back in the form of mucus-bound lamellae that contour the rounded rear end of the echinoid. As in *Nereites*, fecal material is backfilled secondarily as a separate string, but in this case through a sanitary tube formed by longer spines surrounding the anus. In contrast to *Nereites* the fecal string of *Scolicia* is relatively narrow and situated at the base of the whole backfill structure. Also, there may be two anal tufts resulting in two parallel fecal strings (Figure 3-6c–f). Again, the geopetal effect (Figure 3-1) makes the burrow appear very different if preserved on the sole or on the top surface of a sandy layer (Figure 3-6a).

We ignore in this discussion the Paleozoic representatives of the form genus *Scolicia*. Since irregular echinoids originated only in the Jurassic and because the ichnogenus is still a poorly defined holding bag, these burrows must have been produced by other, possibly sluglike organisms that burrowed in a similar conveyor-belt fashion, but with peristaltic waves passing along the soft body. Pending a careful analysis of all *Scolicia* ichnospecies, however, a diagnostic distinction between the two groups is not yet possible.

Scolicia burrows that were probably made by echinoids first appear in shallow marine sandstones of the Upper Jurassic (Figure 3-6b). They continued in this facies until modern times, but without developing systematic search patterns. It is only in the Upper Cretaceous that they make their appearance in the post-turbidite community of the flysch facies. These immigrants had already evolved guided meander programs, but the lobes remain relatively short (Figure 3-6c). In the Eocene, they also appear in the pre-turbidite community, being distinguished from their post-turbidite ancestors by being preserved merely as washed-out casts and by performing more rigid meander programs with a distinct starter spiral (Figures 3-3 and 3-6e). In some cases they

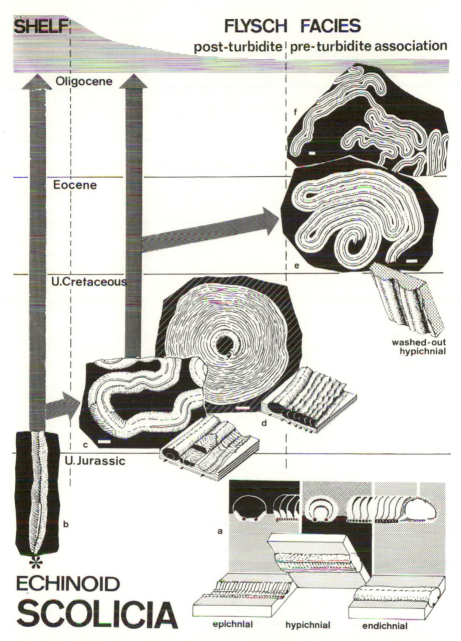

SHELF

FLYSCH FACIES
post-turbidite | pre-turbidite association

Oligocene

Eocene

U.Cretaceous

U.Jurassic

f

e

washed-out
hypichnial

c

d

b

a

ECHINOID
SCOLICIA

epichnial   hypichnial   endichnial

FIG. 3-6 While the origin of Paleozoic representatives of the "holding bag" ichnogenus *Scolicia* is still uncertain, post-Triassic forms with complex backfill structures (**a**) can be related to burrowing irregular echinoids that had their origin in the Jurassic. As in the case of *Nereites* (Figure 3-5), systematic meandering behavior evolved in the deep-sea (flysch facies), where some species also managed to intrude the pre-turbidite association dominated by graphoglyptids (Figures 3-10, 3-11, 3-12). Note also the ability of more advanced species (**f**) to meander without a starter spiral and the systematic violation of the order not to overcross previous turns (**d**) (**b–f** from Seilacher, 1978).

have been found following graphoglyptid burrow systems (Seilacher, 1977, Figure 2b), which might suggest that they also fed on the ecto-symbionts in these "mushroom gardens" (see below). By Oligocene times, the meander lobes of pre-turbidite *Scolicia* forms have become longer and no more need a starter spiral, but can develop from straight, non-guided stretches, during which the animal had "switched off" the meandering behavior. We also observe a gradual miniaturization, which is so common in flysch trace fossils.

Another miniaturized trend of *Scolicia* is represented by dense, coaster-shaped spirals with up to 25 whorls but no meander loops (Figure 3-6d). This form occurs in the post-turbidite association of the Upper Cretaceous and differs from associated feeding burrows by systematically violating the paradigm for sediment feeding: instead of being guided, each whorl cuts deeply into the backfill structure of the previous one, leaving only a remnant of crescentic cross-section. This means that along the spiral path only about one third of the handled sediment was fresh; another third had been processed once and the rest twice before by the same individual. Such violation only makes sense if new food became available with time in the reworked sediment, for instance through fermentation by ectosymbiotic organisms. Once introduced by the echinoid, they would have been propagated like sour dough by the triple reworking. Such an assumption may appear wild, but we should remember that on its long way from the zones of primary production the organic detritus has become very much impoverished in degradable compounds and consists mainly of non-degradable materials by the time it reaches the deep-sea bottom. Utilization of these residues by symbiotic fermentation therefore should be expected in this environment, and in fact we shall see a lot more of it when dealing with the graphoglyptid tunnel systems in a later section.

BEHAVIORAL SEQUENCES IN SPREITE BURROWS

While the previously discussed forms represent feeding tracks (pasci-chnia) of infaunal but vagile sediment feeders, spreite burrows are related to a hemisessile, tube-dwelling mode of life. Such tubes are commonly U-shaped in order to facilitate continuous ventilation. In tubes that have no indurated wall, this U can be enlarged to fit the size of the growing inhabitant by gradually deepening (and widening) the basal bend. The result is a transverse backfill structure called spreite by early paleontologists who held these burrows to be fossilized algal fronds. In the context of this volume, spreite burrows are interesting because they may represent a large enough part of the inhabitant's life time to record ontogenetic changes in behavior.

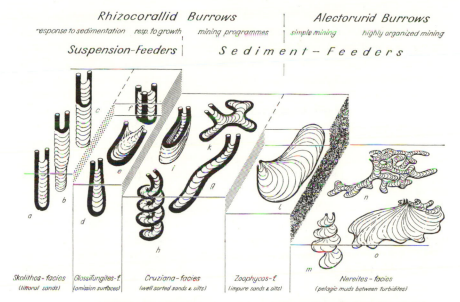

BATHYMETRY of SPREITE BURROWS

FIG. 3-7 Spreite burrows grow by expansion of a U-tube and formation of transversal backfill structure ("spreite"). In the vertical forms of more turbulent shallow marine environments, this expansion adjusts for the growth of the inhabitant and for the change of surface level by sedimentation and erosion (**b, d**). In deeper environments such hair-pin burrows tend to become inclined or horizontal and much longer than body growth would require, indicating that the reworked sediment became systematically processed for food. At the same time burrowing programs become more complex. The figured rhizocoralliid ichnospecies were probably made by shrimp-like crustaceans (see Figure 3-8). Alectorurid burrows, in contrast, are probably worm-made, but they show, in still quieter and deeper environments (the Nereites facies corresponds to deep-sea flysches), to evolve more complex burrowing programs (from Seilacher, 1967a).

Rhizocoralliids

Two major groups of spreite burrows, rhizocoralliid and alectorurid, can be distinguished by morphological and environmental criteria (Figure 3-7). Rhizocoralliid forms, exemplified by the common trace fossil *Rhizocorallium*, have the shape of a hair pin and wide tubes compared to the width of the U. Since burrows of this type are today produced by such different animals as mayfly larvae, amphipod crustaceans, and spionid worms, a heterogeneous nature must also be expected for the fossil representations. But for the forms that are most common in shallow marine deposits, a crustacean origin is indicated by size, the bifid nature of scratches, and the small diameter of fecal pellets that are seen on the walls in particular modes of preservation.

Even in this narrower sense, rhizocoralliid spreite burrows served a variety of biologic purposes. In the simplest case *(Glossifungites)*, the spreite is no more than a by-product of tube enlargement by an organism that got its food by suspension feeding or browsing from outside the sediment. In vertical rhizocoralliid burrows *(Diplocraterion)* active displacement of the basal bend may also be induced by sedimentation or erosion at the surface, resulting in either a retrusive or protrusive spreite structure (Figure 3-8) or in a combination of both *(Diplocraterion yoyo;* Goldring, 1964).

In less turbulent environments, where sedimentation rates are lower and more organic detritus can accumulate in the sediment, the spreite burrow technique can be also used for sediment feeding. The resulting burrows are always protrusive, but in an oblique or horizontal rather than a vertical direction. They also tend to become much longer than would be necessary to accommodate growth and commonly contain a high percentage of fecal pellets in the spreite. Such feeding burrows (fodinichnia) may modify the simple hair-pin geometry into lobate, helicoid, or composite forms (Figure 3-7), but they never evolved continuous strip mining by a composite spreite.

In order to understand this limitation, it is necessary to analyze the monitoring mechanism of rhizocoralliid burrows. A basic clue lies in the fact that the limb distance remains the same throughout the spreite except for the growth increase. It is also telling that in modern examples the makers are rather elongate, with their length equalling or exceeding limb distance. In vertical U-tubes this could simply mean that the animal burrows along its whole length and only when it lies horizontally in the deepest part of the burrow. In horizontal U-burrows, however, gravitational monitoring fails, so that the bend of the body is the only available trigger. The validity of this interpretation is shown by lobate systems (Figure 3-8d) in which accidental bends in the limbs of the original U induced the construction of new spreite lobes. In one case (Figure 3-8e), a concave bend has even made the animal burrow in the wrong direction, cutting into the previous spreite and almost intersecting with the tunnel on the other side.

Also in agreement with this behavioral model is a slipper-shaped ichnospecies of *Rhizocorallium* from the Upper Jurassic (Figure 3-7i), in which production of an initial inclined spreite tongue is followed by a retrusive upward displacement of the whole U-tube. In terms of behavior this means that the ordinary protrusive spreite program stopped once a certain size and depth had been reached. From then on the tube simply moved up by sediment being scraped from the roof and deposited at the tunnel floor.

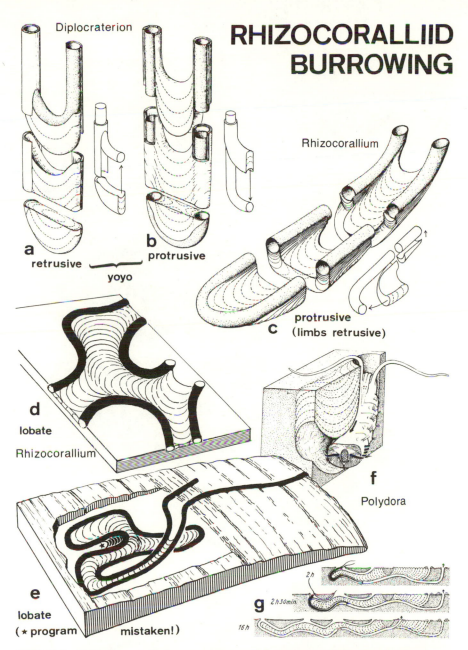

RHIZOCORALLIID BURROWING

Diplocraterion

a retrusive } yoyo

b protrusive

Rhizocorallium

c protrusive (limbs retrusive)

d lobate Rhizocorallium

e lobate (* program mistaken!)

f Polydora

g 2 h | 2 h 30 min. | 16 h

FIG. 3-8 Whether made by crustaceans (ichnogenera *Diplocraterion* and *Rhizocorallium*) or by modern spinoid worms (*Polydora*), rhizocoralliid burrows have in common a hair-pin outline of the inhabited tube. The configuration of lobate versions and the "mistake" in a *Polydora* boring in a modern oyster shell (**e**), which almost led to the interruption of the opposite tube limb, show that it is a critical bend in the tube that induces the animal to proceed with the extension of the spreite (a–c from Seilacher, 1967a; f–g from an experiment with *Polydora ciliata* burrowing in a mud-filled petri dish).

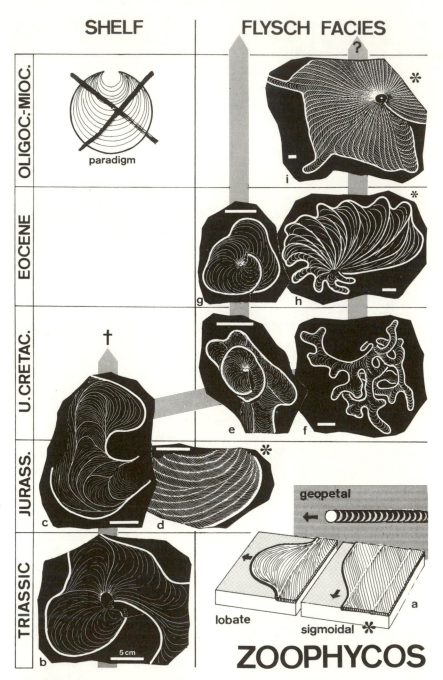

FIG. 3-9 In contrast to rhizocoralliid burrows, the (unknown, but probably worm-like) producer of *Zoophycos* does not need a certain bend of the marginal tube to perform the spreite production. More highly organized forms of the flysch facies, however, have restricted this activity to defined stretches, which resulted in the formation either of distinct lobes or tight spirals, or to sigmoidal

Alectorurids

This group appears to be more homogeneous in a taxonomic as well as ecologic sense. The main ichnogenus is *Zoophycos*, known since the Ordovician and common in modern deep-sea sediments (Wetzel and Werner, 1981). The makers of *Zoophycos* are as yet unknown, but behavior patterns and the large size of the fecal pellets relative to tunnel diameter suggest that it is a worm-like organism.

In contrast to rhizocoralliid burrows, the marginal tunnel of *Zoophycos* is narrow compared to the width of the U, so that the spreite forms a relatively thin body and appears in vertical sections as a horizontal band with crescentic backfill lamellae. Since there are no vertical or retrusive modifications, we can assume that all *Zoophycos* burrows served for sediment feeding. This general homogeneity encourages us to interpret the fossil record of *Zoophycos* in terms of behavioral evolution.

At this point, it is important to emphasize the behavioral difference between rhizocoralliid and alectorurid burrows. Rhizocoralliid burrowing is initiated and monitored by a critical bend or curving radius. It therefore needs a U-tube for a start and can progress only in separated lobes. *Zoophycos*, in contrast, can start spreite lobes burrowing from an almost straight tube (Figure 3-9a), reducing the turning radius during the process. But it can also proceed along a sigmoidal bend of the active section (Figure 3-9a). This versatility has allowed *Zoophycos* to evolve a variety of complex spreite patterns, some of which reflect very systematic and effective strip mining (Figures 3-9d, 3-9i).

The wide environmental range of *Zoophycos* provides another advantage for our review. While being generally restricted to low-turbulence environments, in which organic detritus becomes easily settled, it occurs in protected-shelf as well as in deep-sea environments, including the post-turbidite association of flysch sediments. But the distribution

---

bends, which allowed an effective way of strip mining. Note that sigmoidal strip mining succeeds a lobate phase in h and that it proceeds contouring the spiral in d but radially in i, which form spirals upwards and uses different programs to reach the initial depth and to periodically reset the diminishing radius of the whole system. Behavioral constraints, however, have excluded *Zoophycos* from ever reaching the theoretical paradigm, in which the volume of the processed sediment would be maximized relative to the length of the marginal tube (f, h from Frey and Seilacher, 1980; b, Upper Triassic, Steinplatte, Austria, Catalogue of Geolog.-Palaontol. Instit, Tubingen GPIT, 1627/1; c, Bajocian, Wutach, GPIT 1627/2; d, *Zoophycos*-Dogger, Switzerland, GPIT 1627/3; e, Bergheim near Salzburg, GPIT 1627/4; g, Pontassieve near Florence, GPIT 1627/5; i, Amberley Limestone, Montunau River, New Zealand, drawing combined from Lewis, 1970, and field observations; some of the drawings are reversed to show uniform clockwise turning).

pattern of *Zoophycos* also contains an interesting historic element, with the record in shallow marine sediments starting in the Ordovician and ending in the Cretaceous, shortly after the mid-Cretaceous immigration wave had brought *Zoophycos* into deep-sea sediments, where we observe—as in *Nereites* and *Scolicia*—the evolution of more systematic and complex behavior patterns and continuation of the record up to present times.

Leaving out the Paleozoic record, which is again complicated by the unsatisfactory distinction of apparent homeomorphs (*Spirophyton, Daedalus*, etc.), we start our discussion with Triassic forms (Figure 3-9b). They are irregularly lobate, with a tendency to turn into a spiral. The ordinary Jurassic form (*Z. scoparius*, Figure 3-9c) is similar in outline, but adds new spreite sets not only at the arcuate tip but also along the sides of the initial tongue, thereby reducing tunnel length relative to the spreite area. In a slightly deeper, but still sandy facies (*Zoophycos*-Dogger of the Swiss Pre-Alps), however, we find this form associated with another form (Figure 3-9d), in which strip mining regularly proceeds around the spiral circumference with a sigmoidal front.

In the real flysch facies, the *Zoophycos* record starts in the Upper Cretaceous, where by now the evolution of calcareous plankton had made calcitic oozes a predominant sediment type. Of the two *Zoophycos* types that can be observed, the smaller one (Figure 3-9e) descends into the sediment spirally around one limb of the U-tube (probably the posterior one with relation to the position of the animal). Still it does not reach the spiral staircase design of helicoidal rhizocoralliids (*Lapispira*, Figure 3-7h), but rhythmically expands into larger and larger lobes. In the second type (Figures 3-7n and 3-9f), similar lobes radiate far out with a tendency to curve back, while the spiral plan becomes a subordinate pattern. Since available specimens may be incomplete, it is still unclear, however, whether the two forms represent different ichnospecies or two phases in one burrowing program, the first one being designed for the initial penetration, the second for exploitation of a wider area at the adequate tier.

A corresponding pair of forms is found in Eocene rocks of the same facies. The small helicoidal form is now less lobate and has many (up to at least five) whorls (Figure 3-9g). The associated large form (Figures 3-7b and 3-9h) is radially lobate but only during the first phase, after which it switches to strip mining. As in the Jurassic example (Figure 9d), the backfill lamellae within the strips have a sigmoidal strike, but less regular. Also, instead of proceeding protrusively along the spiral contour, these strips grow from the center radially out, that is, retrusively with respect to the spiral and probably to the animal. Therefore strip mining has presumably evolved independently in the two forms.

In evolutionary terms the transition from the lobate Cretaceous to the two-phase Eocene form could be interpreted as a palingenetic shift of ontogenetic behavior programs, very much as we do in morphogenetic programs. Biologically, it meant a considerable gain in efficiency, because in a spreite burrow strip mining not only improves coverage, but at the same time reduces the length of the marginal tube, through which water has to be pumped for ventilation.

This evolutionary optimization continued within Tertiary times. In the Oligocene of New Zealand (Lewis, 1970) and the Miocene of Italy (Bellotti and Valeri, 1978) we find forms in which the radial strips are very regular and do not bulge out at the margin (Figure 3-9i). The advancing front (as recorded by the backfill lamellae) is even more markedly sigmoidal. Its angle becomes increasingly steeper and thereby widens the strip as it advances outward; but since the maximum angle of the sigmoidal front is usually less than 90 degrees when it reaches the margin, the length of the radial strip tends to become smaller, rather than growing, as the spiral proceeds. To counteract this effect, the animal produced at more or less regular intervals distinctive reset lobes that project far beyond the marginal spiral and permit the next strip to be made sigmoidally again, but with adequate length. There were probably additional functions for these lobes (such as probing the neighborhood), because they project much more than needed for resetting and against the restriction to keep the length of the marginal tube as short as possible.

While the homology of the reset lobes with the lobate phase of the Eocene form remains to be discussed, it is clear that the lobate starter program as such has disappeared and been functionally replaced by a tongue-like portion, in which sigmoidal strips contour rather than radiate (Figure 3-9i). But in contrast to the Jurassic form (Figure 3-9d) they are retrusive relative to the progression of the whole spiral. Since this starter tongue is relatively steep, it allows a fast penetration to the ultimate depth, from which radial strip mining can proceed in an outward rather than in the inward spiral of the truly helicoidal programs (Figures 3-9e and 3-9g).

It should also be emphasized that the Oligocene–Miocene burrows are much larger than the earlier types, reaching close to 1 m in diameter. This exception to the miniaturization trend observed in other flysch trace fossils may be related to the fact that the host sediments are chalks without obvious turbidites.

In summary, the biostratigraphic record of *Zoophycos*, although still very spotty, confirms the general rule that sediment feeders evolve more rigorous and more complex burrowing programs as they invade deep-sea bottoms. Whether the program successions observed within the individual burrows represent an ontogenetic pattern or a repeatable procedure is still uncertain. In any case it is tempting to view them in terms of developmental acceleration or retardation, that is, in the same way as we use to describe the evolution of morphogenetic programs (Gould, 1977).

The example of *Zoophycos* also shows the constraints of alectorurid burrowing programs: the theoretical paradigm, in which the U-tube expands with a circular outline and therefore with a minimal length (Figure 3-9), was never approached. Obviously, with all flexibility in turning radius, the length of the working line along the tunnel was the limiting factor.

BEHAVIORAL RADIATION IN GRAPHOGLYPTID BURROWING

The burrows we have discussed so far are members of the *post-turbidite association* of flysch trace fossils. Together with many other forms not covered in this chapter (for an overview, see Seilacher, 1978, Figure 1) they made use of the food supply that was episodically introduced by turbidity currents into marginal deep-sea environments, where it became deposited in a graded fashion, mainly in the upper parts of the individual turbidites. Measured in an animal's lifetime, these events were certainly rare and had a limited regional extent, but on the scale of basins they were common enough to provide a continuous niche for species with adequate possibilities of larval dispersal. While the constituent species had to be opportunistic in this respect, their behavior was nevertheless highly specialized compared to species from the shallow marine realm.

Such comparison is possible because in the post-turbidite association recruitment from the shelf was relatively common and can be traced in several cases. Over geological time, however, this evolutionary immigration seems to have been discontinuous, with a sharp rise in the Upper Cretaceous, when many important elements, including *Scolicia* and *Zoophycos* but also miniaturized shrimp burrows *(Granularia)*, appear for the first time in the flysch facies. Also interesting is the trend for the newcomers to be become established in the lower levels of the tiered sediment-feeder community, where food contents were probably lower and energy costs for burrowing higher than in the upper tiers already occupied.

This tiering is usually well preserved in post-turbidite bioturbation horizons, whose sharp tops permit distinguishing between the lithologically similar turbiditic and hemipelagic parts of the shales in the field. Since the original tiering would have become obscured if the bioturbation had continued during the subsequent hemipelagic mud sedimentation, we can assume that each post-turbidite community lasted for only a short time at a given place.

The hemipelagic muds that accumulate on top of the turbidites appear to be generally devoid of trace fossils. This observation, however, is deceiving. The sole faces of the following sandy turbidites are commonly covered with traces, whose preservation (sometimes fluted) shows that they are pre-turbiditic in origin and that they represent not simple replicas of surface trails, but mud burrows eroded during the initial phase of the turbidity current.

This *pre-turbidite* ichnocoenosis faces us with two major questions: (1) Why is it so distinct from the post-turbidite association without overlap of species in spite of the lithologic similarity of the muds in which the burrows were originally made? (2) Why are the burrows of the pre-turbidite association invisible in the hemipelagic mud and need turbidite erosion to be preserved?

The first question must have to do with the nature of the food brought down from the shelf by turbidity currents compared to the planktonic rain of the intervening periods. The second question hints at a basic difference between the two communities in the way they burrowed and exploited the sediment.

Another difference between the two associations refers to their taxonomic structure: while the post-turbidite association is very heterogeneous, the pre-turbiditic background community consists almost exclusively of one group of burrows called graphoglyptids. Nevertheless its diversity at the ichnospecies level is higher than that of the post-turbidite environment, at least in Upper Cretaceous and Tertiary flysches. This is the result of a drastic diversity increase probably by mid-Cretaceous times. It coincides with the diversity burst in the post-turbidite association, although it reflects within-biotope radiation of behavior patterns rather than a wave of evolutionary immigration.

In order to understand the meaning of this behavioral radiation, we must have a closer look at the known types of graphoglyptid burrow systems (Figures 3-10, 3-11, 3-12). At a first glance their patterns might appear to correspond to the paradigm of surface utilization by sediment feeders. But in fact they violate this paradigm in important respects:

1. The unbranched meanders (Figure 3-10a) leave much space unused between their first- and second-order turns. The occurrence of shortcuts further demonstrates that the burrows were open tunnels for repeated usage rather than backfill structures.

2. In the branching and net-shaped systems the taboo of double coverage had to be systematically violated already in the original production.

3. Both branching meanders and nets evolve multiple exits that would be of little help for active ventilation of the system.

4. Most graphoglyptid burrow patterns appear to be highly overcomplicated and difficult to interpret in terms of improved space utilization under the sediment feeder paradigm.

Because of these discrepancies it has been suggested (Seilacher, 1977) that graphoglyptid burrows were not made for direct sediment feeding but to cultivate microorganisms. Such "mushroom gardens"

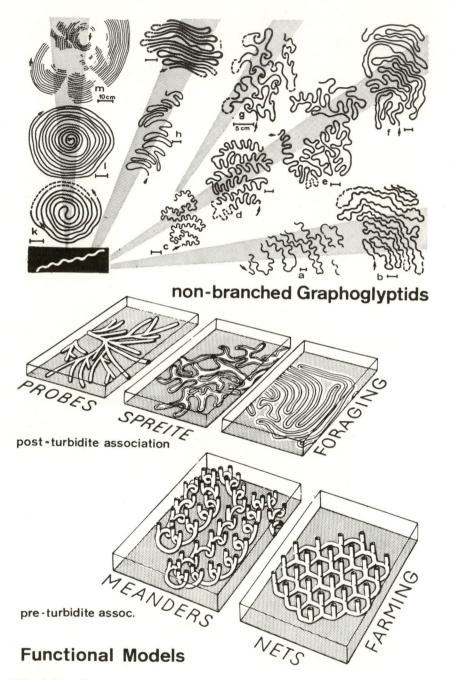

non-branched Graphoglyptids

post-turbidite association

PROBES SPREITE FORAGING

pre-turbidite assoc.

MEANDERS NETS FARMING

Functional Models

FIG. 3-10 Graphoglyptid burrows are preserved only as erosional casts on turbidite soles, so that observed patterns reflect only two-dimensional patterns of potentially more complex three-dimensional systems. But even in this expression and in their simplest, non-branching version they appear overly

would be able at the same time to utilize food from the unreworked sediment around the tunnels and to break down non-refractable substances that a metazoan host is unable to digest. In the mushroom garden model the paradigmatic pattern would be a meshwork rather than a tight meander; the multiple exits would help diffusive ventilation and the interspecific pattern modifications would become safety keys to avoid usurpation of the gardens by other species while the owner was absent or in other parts of the burrow system.

This hypothesis is testable on more than just the grounds of functional morphology. Since the presence of typical graphoglyptid tunnel systems in modern deep-sea sediments has recently been demonstrated by bottom photography (Rona and Merill, 1978), it is only a question of logistics and time when adequate samples will be available to make the critical microbiological tests.

CONCLUSIONS

While the burrowing behavior of marine invertebrates as reflected by trace fossils is obviously far removed from the behavior of terrestrial animals treated in most other contributions to this volume, it can be similarly analyzed in terms of homology and analogy and in addition allows us to trace behavioral evolution through hundreds of millions of years.

Because of the general trend of burrowing programs to become more rigorous and more complicated in deep-sea environments, the most revealing evidence for such evolution comes from flysch trace fossils. They suggest that behavioral patterns evolved much like morphological features, being directed by functional optimization, character displacement, ontogenetic acceleration and retardation, and in some cases a switch in the functional paradigm.

---

complicated and diversified for the simple purpose of effectively scanning a given area. This is all the more surprising in view of the demonstrable fact that the areas between the tunnels have not been touched by the animal. The functional model suggests that the deviation from the meander burrow paradigm of sediment feeders (as exemplified by nereitids; see Figure 3-5h) is in graphoglyptids replaced by another one, in which the areas between tunnels are equalized for exploitation by other means. This could be done by attraction and trapping of mobile infaunal organisms (like in the modern worm *Paraonis;* see Papentin and Roder, 1975) or by the association with microorganisms (fungi or bacteria) that are able to extract nutrients from the sediment without burrowing ("farming"). The patterns of more complex graphoglyptid tunnel systems (Figures 3-11 and 3-12) speak for the second possibility (from Seilacher, 1977).

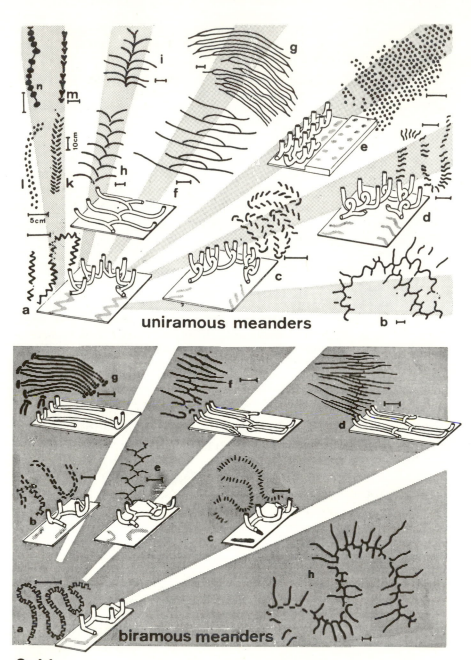

uniramous meanders

biramous meanders

3-11

FIGS. 3-11 and 3-12 Branching graphoglyptid tunnel systems violate the paradigm of efficient sediment feeding not only by the double coverage that is necessary to produce branches and anastomoses, but also by the introduction of many accessory openings. These forms support the "mushroom garden"

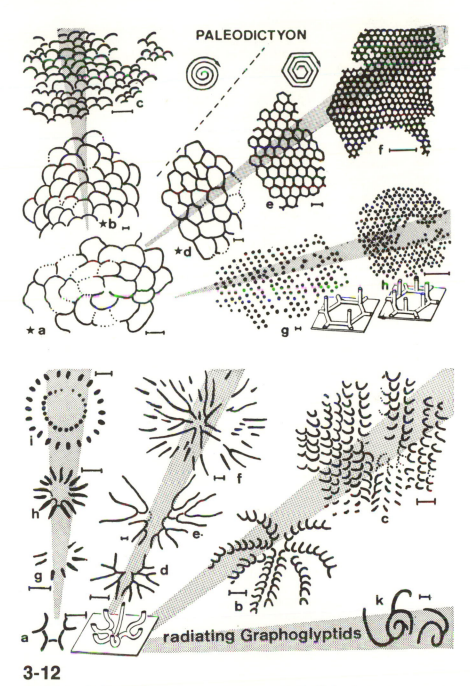

PALEODICTYON

radiating Graphoglyptids

3-12

model, in which modification of basic programs into distinctive patterns would increase fitness not by optimization, but as a safety key against usurpation by other species.

Trace fossils also show the response of evolution to long-term changes in the biosphere. But in contrast to shallow marine environments, in which extinctions and following radiations were a major driving force, community evolution in the deep sea was a slow, accumulative process, insensitive to the faunal turnovers at the era boundaries. Instead the diversity buildup shows a marked step by Upper Cretaceous times. It can be observed in two communities (pre- and post-turbidite associations) that differ basically with respect to taxonomic spectrum, trophic structure, and evolutionary recruitment. This diversity explosion can be tentatively related to changes in the zones of primary production, that is, in the phytoplankton and the coming of angiosperms (including eelgrass), which increased the influx of non-degradable detritus and thereby induced the development of novel feeding strategies. It is probably more at the level of such major patterns than by the specific details that fossil behavior can contribute to our understanding of evolutionary processes.

LITERATURE CITED

Bellotti, P., and P. Valeri. 1978. L'influenza dell'ambiente sedimentario sull'assetto elicoidale delle strutture a *Zoophycos*. *Bollettino della Societa Geologica Italiana* 97:675–685.

Chamberlain, K. C. 1978. A guidebook to the trace fossils and paleoecology of the Ouachita geosyncline. Society of Economic Paleontologists and Mineralogists, pp. 1–68.

Fedonkin, M. A. 1981. White sea biota of the Vendian. Precambrian nonskeletal fauna of the northern Russian platform. *Transactions Akademiia nauk Soviet Socialist Republic* 342:1–100.

Frey, R. W., and A. Seilacher. 1980. Uniformity in marine invertebrate ichnology. *Lethaia* 13:183–207.

Glaessner, M. F. 1969. Trace fossils from the Precambrian and basal Cambrian *Lethaia* 2:369–393.

Goldring, R. 1964. Trace fossils and the sedimentary surface in shallow-water marine sediments. In: Van Straaten, L. M. J. U. (ed), Deltaic and shallow marine deposits. *Developments in Sedimentology* 1:136–143.

Goldring, R., and A. Seilacher. 1971. Limulid undertracks and their sedimentological implications. *Neues Jahrbuch für Geologie und Palaontologie Abhandlungen* 137:422–442.

Gould, S. J. 1977. Ontogeny and Phylogeny. Harvard University Press, Cambridge, Mass.

Lewis, D. W. 1970. The New Zealand *Zoophycos*. *New Zealand Journal of Geology and Geophysics* 13:295–315.

Papentin, F., and H. Röder. 1975. Feeding patterns: The evolution of a problem and a problem of evolution. *Neues Jahrbuch für Geologie und Palaontologie Monatshefte* 1975:184–191.

Raup, D. M., and A. Seilacher. 1969. Fossil foraging behavior: Computer simulations. *Science* 166:994–995.

Rona, P. A., and G. F. Merill. 1978. A benthic invertebrate from the mid-Atlantic ridge. *Bulletin of Marine Science* 28:371–375.

Seilacher, A. 1960. Lebensspuren als Leitfossilien. *Geologische Rundschau* 49:41–50.

Seilacher, A. 1967a. Bathymetry of trace fossils. *Marine Geology* 5:413–428.

Seilacher, A. 1967b. Fossil behavior. *Scientific American* 217:72–80.

Seilacher, A. 1977. Pattern analysis of Paleodictyon and related trace fossils. In: Crimes, T. P., and J. C. Harper (eds), Trace fossils 2. *Geological Journal* (Liverpool) Special Issue 9:289–334.

Seilacher, A. 1978. Evolution of trace fossil communities in deep sea. In: Seilacher, A., and F. Westphal (eds), Paleoecology. Constructions, sedimentology, diagenesis and association of fossils. *Neues Jahrbuch für Geologie und Palaontologie Abhandlungen* 157:251–255.

Wetzel, A., and F. Werner. 1981. Morphology and ecological significance of *Zoophycos* in deep-sea sediments off NW Africa. *Palaeogeography, Palaeoclimatology, Palaeoecology* 32:185–212.

4

The Evolution of Predator-Prey Behavior: Naticid Gastropods and Their Molluscan Prey

Jennifer A. Kitchell

THE fossil record rarely chronicles evidence sufficient to assess the historical evolution of behavior (Simpson, 1958; Brown, 1975; Hinde, 1982). Examples of fossilized behaviors frequently represent catastrophic preservation processes or frozen moments in time much like Pompeii's mummified citizenry. Exceptions are those instances in which behavior characteristically fossilizes. One such exception is represented by the behavioral artifacts of foraging behavior. The foraging behaviors of deposit-feeding marine organisms are reasonably well archived in the fossil record and are discussed by Seilacher in Chapter 3 of the present volume. In this instance, however, the distribution of food resources within the sediment, a primary control on foraging behavior, is generally not known (see Kitchell, 1979). By contrast, the second exception involves the more addressable problem of the simultaneous evolution of organism and (biotic) environment, and is the topic of this chapter.

Drilling marine gastropods of the family Naticidae and their molluscan prey provide one of the few cases of sufficient direct evidence for the study of behavioral interactions over an evolutionary time scale (Figure 4-1). The fossil evidence of naticid predation spans more than 100 million years (Sohl, 1969; Taylor et al., 1980; see Fursich and Jablonski, 1984, for naticid-like boreholes of Triassic age) and bears witness to two modes of evolutionary change, behavioral and morphological. The naticid predator marks each prey with a readily fossilized morphological signature of its behavioral strategy. A beveled borehole provides tangible evidence of whether or not predation was attempted, whether it succeeded or failed (Kitchell et al., in press), at what site, and an estimate of the predator's size (Kitchell et al., 1981). Because

RETRIEVABLE INFORMATION

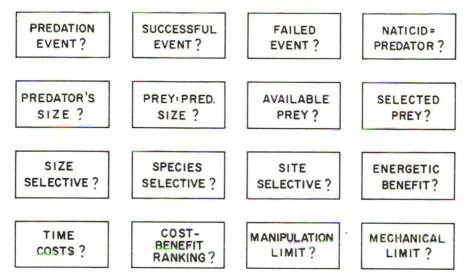

| PREDATION EVENT ? | SUCCESSFUL EVENT ? | FAILED EVENT ? | NATICID = PREDATOR ? |

| PREDATOR'S SIZE ? | PREY:PRED. SIZE ? | AVAILABLE PREY ? | SELECTED PREY? |

| SIZE SELECTIVE ? | SPECIES SELECTIVE ? | SITE SELECTIVE ? | ENERGETIC BENEFIT? |

| TIME COSTS ? | COST-BENEFIT RANKING? | MANIPULATION LIMIT ? | MECHANICAL LIMIT ? |

FIG. 4-1 Summary of the retrievable information content of the naticid predatory-prey interaction. The interaction provides morphological evidence of the success or failure of a predation event, the identity and size of the predator, the prey:predator size ratio, the size and species distribution of available vs. selected prey, the degree of predatory behavioral selectivity for prey species, prey size classes, boring site, an estimate of the cost (drilling time)/(energetic) benefit ratio per prey, and predatory constraints (mechanical and manipulation).

these data are recorded directly on the hard-part morphology of the prey, an associated measure of prey characteristics per predation event is archived.

Predators with flexible behavioral repertoires may evolve under rules for evaluating and selecting prey based on learning as a consequence of individual experiences. The studies, discussed below, indicate that short-term learning behavior, based on individual experiences with prey, is less important in this interaction than long-term stereotypic responses, suggesting that the evolutionary dynamics of prey selection behaviors may be studied appropriately using the integrated evidence derived from fossil assemblages.

Although the interaction between naticid predator and molluscan prey is coupled, the essential component of such a predator-prey interaction is conflict of interest, making the problem not one of simple maximization. Moreover, both predator and prey may evolve independently and at different rates. This chapter reviews the empirical evidence that naticid predators behaviorally select prey, evaluates a model of prey selection to account for both present-day behavioral interactions and the evidence of behavioral interactions in the fossil record, describes the consequences of feedback on behavioral energetics and assesses the potential for coevolution, and presents evidence of inefficient components of predatory behavior that persist over spans of evolutionary time.

Despite substantial diversification and taxonomic turnover in this predator-prey system as documented by its evolutionary history, the stereotypy of rules employed in prey choice appears robust. Since their first well-documented appearance in the fossil record, during the Late Cretaceous, predatory (drilling) naticids seem to have followed the same general principles with regard to site selection on individual prey, size selection within species, and prey selection. Deviations from these rules offer the greatest potential for insight into historical constraints or alternate selection pressures.

WHAT GOVERNS THE BEHAVIORAL SELECTION OF PREY?

An understanding of the factors governing the behavioral selection of prey is necessary to test predictions of either differential rates of evolution or reciprocal evolutionary responses between predator and prey. The behavioral selection of prey by naticids had previously been described as indiscriminate (e.g., Hoffman et al., 1974; Stanton and Nelson, 1980). We initiated a series of studies (Kitchell et al., 1981) to determine whether prey selection by predatory naticid gastropods is random or, instead, the behavioral selection of prey maximizes net energy return to the predator.

This series of studies addresses questions of behavioral strategies and their evolutionary consequences. It is assumed that behaviors of prey selection have heritable variation and that naticid gastropods experience both historical and architectural constraints.

Naticid predators are known to exhibit strong prey preferences. In particular, several studies (e.g., Edwards, 1974; Edwards and Huebner, 1977; Wiltse, 1978) have determined the preference ranking of prey species for a typical naticid, *Polinices duplicatus*. Subsequent choice experiments revealed that these preference rankings could not be easily extinguished by training on alternate prey. Regardless of training diet, the predators switched to selecting the originally preferred species when provided a choice. Our studies were designed to implement the hypothesis of nonrandom prey selection by determining prey handling times by *P. duplicatus* for those prey species with known preference rankings and, in addition, to determine rate of boring, change in borehole geometry as a function of boring time, and mechanical and manipulation limits to successful predation.

A model was formulated in which net energy return to the predator is maximized when the behavioral selection of prey proceeds in descending rank order of species-specific cost-benefit functions. Such functions are determined by four species-specific empirical parameters (Kitchell et al., 1981): energetic value of the prey item, the probability of successful predation, prey handling time defined as drilling time, and prey recognition time. A fifth parameter, prey availability, does not affect the ranking of prey but determines diet breadth. If the hypothesis of energy maximization correctly predicts the behavioral selection of prey, then predicted rankings of prey derived from cost-benefit functions should correspond to actual prey preference rankings.

The results indicated that prey selection, based on chemical sensing and tactile evaluation, is not random with respect to either prey species or prey size. Prey selection, both within and between species, was found to be consistent with the hypothesis of a behavioral strategy of energy maximization (Kitchell et al., 1981). Prey selection behavior represented a predictable response to multiple prey parameters. Such results provide a potential means of assessing the evolutionary record of prey selection behavior.

VERIFICATION OF THE ENERGY MAXIMIZATION ASSUMPTION AND A TEST FOR STEREOTYPY

The energy maximization hypothesis makes the independent prediction that handling time limits the number of prey. Consequently, a second test of the validity of the energy maximization argument follows: if handling time is reduced, that time should be utilized to handle additional prey.

Drilling time is directly related to shell thickness of the prey (Kitchell et al., 1981); this prediction can thus be tested by altering shell thickness. If the predator's behavioral strategy of prey selection is one that maximizes net energy gain (reproductive success is dependent on surplus energy; Huebner and Edwards, 1981), then a reduction in drilling time should result in the selection of additional prey during a fixed interval of time. By contrast, if the predator's behavioral strategy is not to increase the total amount of energy acquired, but rather to increase the amount of time available for other activities (e.g., time minimization strategy; Schoener, 1971), then the realized reduction in handling time should not affect predation rate.

The experimental procedure (Boggs et al., 1984) involved providing the predator with a novel prey type. To maintain ceteris paribus except for drilling time, the shell thicknesses of individuals of *Mercenaria mercenaria* were reduced about 50% by grinding. As a consequence, neither energetic value, escape behavior, olfactory signals, size, or general geometry of the "novel" prey type differed from similarly sized "normal" individuals of *M. mercenaria*. Handling time, however, differed substantially: the prey used in these experimental tests averaged 43 mm in size, normally requiring individuals an average of 63 hr of drilling time. Experimentally treated prey, on average, required half the amount of drilling time (consumption time was the same for both normal and experimentally treated prey).

The experimental treatment of grinding also provided the predator with a tactile cue that could be used to discriminate between the two classes of prey with substantially different characteristic handling times. Consequently, the experimental design enabled us to test a second hypothesis. Is prey selection behavior stereotyped (cued, for example, to a species-specific olfactory cue) or will the predators evaluate and select prey on the basis of individual experiences? Both specialists and generalists may be either stereotypic or plastic in their behavioral responses to resources. Morse (1980) defined stereotypy as the behavioral tendency to exploit resources similarly regardless of changing conditions whereas plasticity represents the tendency to alter exploitation behaviors in response to changing resource conditions.

Because we provided a novel prey type with an improved cost-benefit ratio (via reduced drilling time), we provided a situation of choice. Predators could evaluate prey on the basis of individual experiences and switch their preference to the novel prey type. If, on the other hand, the predators did not learn to prefer the novel prey type but chose individual prey at random, then these results would refute the hypothesis that the behavioral selection of prey is based on individual experience.

Results of these experiments (Boggs et al., 1984) were twofold: (1) in all cases, predation rate when only the novel thin-shelled prey were available was greater than on the normal prey; naticid predators did not forage at a fixed rate regardless of prey value and naticid predators did not behave as time minimizers (see Menge, 1974); instead, naticid predators behaved as energy maximizers, utilizing the time saved in drilling the novel prey type to select and process additional prey; (2) there was no consistent evidence of learning to select the novel prey type; prey selection in the presence of both prey types was random, indicating that the preferences of naticid predators are stereotyped and not based on individual experience.

The important implication of these results is that prey preferences represent long-term responses that can be effectively studied using fossil assemblages that represent temporally integrated material. Such stereotypy may represent an appropriate response to a (prey) environment that is unpredictable over a short-time scale but whose long-term mean is predictable (e.g., Heiner, 1983). In this case, the complexity of accurately evaluating each individual prey may prevent the predator

from developing a behavioral strategy of prey selection based on individual experience. Instead, the predator's behavioral strategy depends
on long-term species recognition and the evaluation of size.

STEREOTYPY OF BEHAVIORAL PATTERNS

Size-Selective Behavior

The model of naticid prey selection developed here predicts size-selective behavior. The causal explanation of size selectivity resides in the
shape of the prey cost-benefit function. These functions decrease with
increasing prey size (Kitchell et al., 1981). For a given prey species, the
larger the size class the lower its cost-benefit ratio and the greater the
net energetic return to the predator. Consequently, a population-level
selectivity of prey based on size is expected from the selective behavior
of an age-structured predatory population, provided a sufficiently wide
range of size classes is available. Static cost-benefit functions, derived
solely from prey parameters, however, ignore predator size and have
no inflection point (see Kitchell et al., 1981). The parameter that fixes
the inflection point for a given predator and prey species size is defined
by the probability of successful predation and depends on the
prey:predator size ratio. Hence, although size-selective predatory
behavior is predicted from the energy maximization hypothesis, a size
"refuge" for prey may also exist, in which prey are too large to attack
successfully. Both such population-level phenomena—size-selective
behavior and a size refuge—may be recovered from fossil assemblages
with the caveat that the assemblage represents a time-integrated
sample.

Data derived from fossil assemblages provide empirical evidence that
size selectivity has been a behavioral feature of naticid predation
throughout the evolutionary history of this group (e.g., Ansell, 1960;
Kitchell et al., 1981; Hoffman and Martinell, 1984; Kitchell et al., in
press). Naticid predators were size-selective in the Late Cretaceous, as
illustrated in Figure 4-2, and in all subsequent assemblages we have
examined (an example is provided in Figure 4-3) of sufficient sample
size, a remarkable finding given the time-integrated nature of these
assemblages. Such data indicate that the pattern of size selectivity is
sufficiently robust to be observed despite time integration. (Size selectivity is not expected to characterize all cases due to insufficient sample
sizes, lack of a sufficiently wide range of size classes in the predator's
environment, or the blurring effects of time integration.)

Hoffman and Martinell (1984) recently proposed to test the predictions of the energetics maximization model of Kitchell et al. (1981) by
assessing whether naticid predation was size-selective in a highly diverse
Pliocene assemblage. Size selectivity was found to be statistically signif-

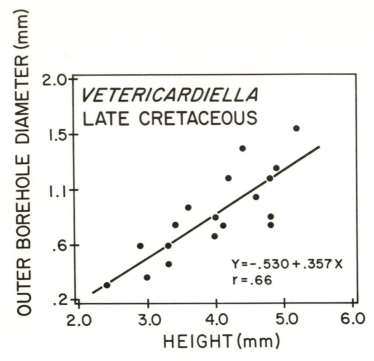

FIG. 4-2 Size-selective behavior in a Late Cretaceous species of *Vetericardiella* from the Ripley Formation. Borehole diameter provides an excellent estimate of predator size (see Kitchell et al., 1981). Predator size and prey size (height) are positively correlated at a significance level of .01.

icant, despite the time averaging of the assemblage. Six species of prey documented size selectivity at the 0.01 significance level (sample size ranged from 27–110). The only species that did not demonstrate size selectivity was represented by a sample size of 10. Studies of modern naticids have similarly demonstrated size selectivity in the intraspecific selection of prey (Franz, 1977; Edwards and Huebner, 1977; Kitchell et al., 1981; Griffiths, 1981; Broom, 1982; Berry, 1983).

The phenomenon of size selectivity also resolves a previously reported paradox. Thomas (1976) had reported highly variable selection of size classes in species of the bivalve *Glycymeris* from three Miocene assemblages. Reanalysis of Thomas's collections, however, revealed that if predator size is included as a variable (derived from measuring borehole diameters), then the variability in selection of size classes is consistent with the size-selectivity hypothesis as expected (Kitchell et al., 1981). Predator size and prey size in these three assemblages are positively correlated. The variability reported by Thomas is the result not of variability in predatory behavior but in predator sizes: two of the assemblages are dominated by small-sized naticids and the other assemblage is characterized by large-sized naticids. The small-sized naticids predictably selected smaller size classes of prey than the large-sized naticids.

The naticid predator-prey system is particularly amenable to providing data on the historical evolution of size-selective behavior. Both the slope of intraspecific selectivity and placement of the inflection point or critical escape size can be quantified. One might expect, for example, the slope of size selectivity to have decreased over evolutionary time as the predatory ability to subdue prey increased. There seems, however, to be little empirical evidence in support of a temporal directionality in the slope of size selectivity. Instead, such slopes vary depending on the size structure and diversity of the predatory population and the size structure and diversity of the prey population.

A confounding problem resides in the fact that size selectivity is a parameter under the joint control of predator and prey. If only the predator evolved, then an increase in predatory efficiency would be manifested as a temporal decrease in the predator:prey size ratio. The empirical prediction would be increasingly smaller boreholes on prey

FIG. 4-3   Evidence of size-selective behavior in an Early Miocene species of *Anadara* from the Chipola Formation (significance level .01).

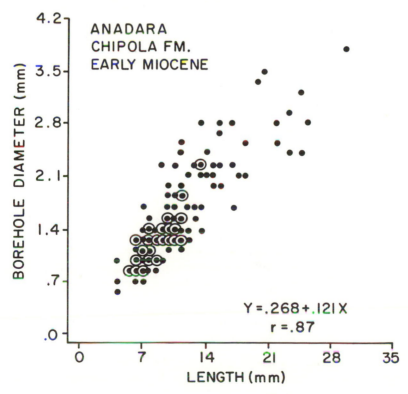

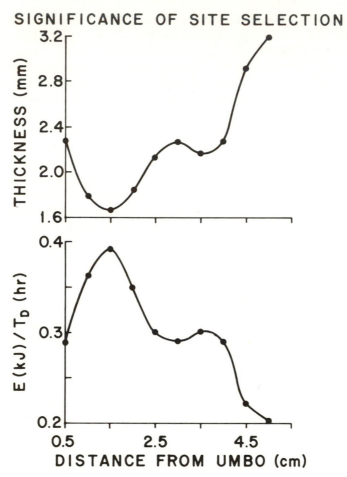

FIG. 4-4 Variability in shell thickness along a transect from the umbo to the valve margin in *Mercenaria mercenaria* (upper panel), and resultant change in potential net energetic return (kJ/hr) (lower panel).

of the same size or a decrease in the slope of borehole size vs. prey size. If only the prey evolved, however, the prediction of the direction of change with time is reversed. The predator:prey size ratio would be an increasing function with time. In reality, the parameter of size selectivity is under joint control of predator and prey and, consequently, could yield the information of no temporal change despite an increase in both predatory and prey efficiency.

Site-Selection or Prey-Handling Behavior

By contrast, variability in siting of the borehole on a particular prey species is a behavioral parameter under the sole control of the predator. Again, site specificity is predicted by the energy maximization hypothesis. Thickness of the prey's shell at the site of boring directly determines drilling time. The relationship between shell thickness and prey size is a species-specific relationship that may be either linear or nonlinear (e.g., Figure 3, Kitchell et al., 1981). Because shell thickness varies over the surface of the shell whereas biomass remains a constant, net energy return per unit time drilling will be affected by the predator's behavior of site selection, as illustrated for *M. mercenaria* (Figure 4-4).

Simulation trials of random siting (upper panel, Figure 4-5), using *M. mercenaria*, resulted in both a lower mean net energetic return and a higher variance of energy per unit time, as compared with the actual range of siting positions on this prey type (middle panel, Figure 4-5). Analysis of actual siting positions for different individual predators on

FIG. 4-5 Comparison of net energetic return as a function of the predator's behavior of site selection, from simulated random siting ($N = 1300$) (upper panel), measures of observed siting ($N = 100$) (middle panel), and a hypothetical optimum siting behavior (lower panel).

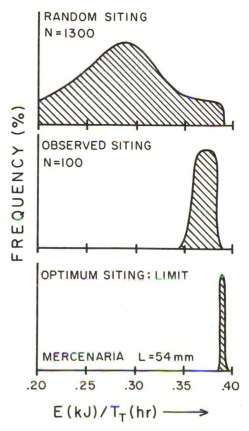

M. mercenaria prey indicates that site specificity is not optimum (see lower panel, Figure 4-5), but the mean net energetic return is substantially higher than random siting behavior and the variance even among individuals is low. Expected costs to the predator have been relatively standardized. From numerous observations of predation events, we conclude that such standardization is the result of a fairly fixed behavioral pattern of manipulating and orienting the prey (Kitchell et al., 1981). Differences among prey species in the characteristic position of the borehole result from differences in the shape of the prey species. Variability in the position of boreholes within a prey species reflects, to a large degree, variability in the sizes of predators on a prey size class. Enveloping prey in the mesopodium of either unusually large or small predators results in abnormal placement of the drilling apparatus. Because predators are behaviorally size selective, this variability is low, provided there is a sufficient availability of prey size classes in the predator's environment.

Although a macroevolutionary trend within the naticid guild of decreased variability in siting of the borehole might be predicted, we have not analyzed our data to determine whether or not there are any significant temporal trends in the evolution of borehole siting. The wide-ranging position of boreholes in species of *Corbula, Caestocorbula*, and *Vetericardiella* from the Late Cretaceous Ripley Formation might be used to support an interpretation of increased site specificity since the Late Cretaceous. Without a rigorous analysis, however, the question remains open.

In this light, an interesting series of studies by Berg has examined quantitatively the degree of stereotypy in borehole siting (Berg and Porter, 1974; Berg and Nishenko, 1975), the effects of maturation on siting behavior (Berg, 1975), and ontogenetic changes in borehole positioning (Berg, 1976). An analysis of the distribution pattern of boreholes evidenced no evolutionary trend (Berg, 1978), and Berg concluded that such behavioral patterns are conservative.

Behavioral vs. Morphological Change

Behavioral strategies may range widely, as Parker (1982) suggested, "without the cost of morphological change." Macroevolutionary morphological trends associated with increasing predatory efficiency are not manifest in our studies of fossil naticid morphology. Increase in the relative size of the foot would presumably enable successful predation on larger prey and promote locomotion. An increase in overall globosity of the shell has been suggested to aid in anchoring the shell within the sediment thereby providing greater locomotory force (Trueman, 1975). To test these predictions, both apertural area (reflecting relative foot size) and overall cross-sectional area of the shell (reflecting globosity, defined as increasing when the cross-sectional area, standard-

TABLE 4-1 Summary of Pacific coast naticid species and localities used to assess predatory morphological change. See Marincovich (1977) for inferred phylogenetic relationships. The allometric relationships of both apertural area and total cross-sectional area, quantified by digitizing standardized photographs of a range of sizes per species, were examined for temporal trends.

Species	Formation	Age
Polinices clementensis	Keasey Fm.	Eocene
P. galianoi	Astoria Fm.	Miocene
P. halli	Ripley Fm.	Cretaceous
P. hornii	Cowlitz Fm.	Eocene
P. hotsoni	Cowlitz Fm.	Eocene
P. lewisii	Elk River Fm.	Pleistocene
P. lewisii	San Pedro Quad, CA	Pleistocene
P. lincolnensis	Lincoln Creek Fm.	Oligocene
P. nuciformis	Cowlitz Fm.	Eocene
P. nuciformis	Astoria Fm.	Miocene
P. pallidus	Elk River Fm.	Pleistocene
P. panameaensis	Viqui Point Fm.	Holocene
P. susanaensis	Lodo Fm.	Paleocene
P. uber	Bahia San Luis Gonzaga	Holocene
Natica janthostoma	Montesano Fm.	Pliocene
N. anthostoma	Astoria Fm.	Miocene
N. clausa	Elk River Fm.	Pleistocene
N. kanakoffi	Astoria Fm.	Miocene
N. posuncula	Temblor Fm.	Miocene
N. rosensis	La Jolla Group	Eocene
N. teglandi	Astoria Fm.	Miocene
N. teglandi	Blakely Fm.	Oligocene
N. uvasana	Tejon Fm.	Eocene
N. weaveri	Lincoln Creek Fm.	Oligocene
Neverita globosa	Cowlitz Fm.	Eocene
N. jamesae	Round Mt. Silt Fm.	Miocene
N. reclusiana	Monterey Bay, CA	Holocene
N. reclusiana	San Pedro Quad, CA	Pleistocene
N. reclusiana	Cowlitz Fm.	Eocene
N. nana	Santa Barbara, CA	Holocene
N. washingtonensis	Lincoln Creek Fm.	Oligocene
N. lamonae	Central Oregon	Holocene

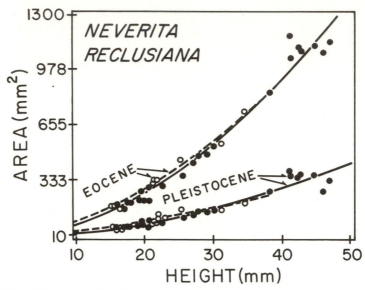

FIG. 4-6 Allometry of total cross-sectional area (globosity, upper slopes) and apertural area (lower slopes) for an Eocene (dashed line) vs. Pleistocene (solid line) assemblages of the naticid species *Neverita reclusiana*. There are no significant temporal differences.

ized for size, increases in response to shape change) were measured over a range of sizes for species of *Polinices, Natica* and *Neverita* (Table 4-1). Neither the allometry of apertural area nor that of cross-sectional area, however, was found to exhibit statistically significant temporal trends. Such morphological stasis is found both within species over time (Figure 4-6) and within lineages over time (Figure 4-7), with the exception of size change as discussed below.

Cannibalistic Behavior

Cannibalism is very common among modern naticids and may be important in population regulation. Cannibalism represents selection of a prey item with a high net energetic return (drilling time is short) but a narrowly available size range (because of active defense; see Kitchell et al., 1981). More indirect effects of cannibalism include the reduction of competitors and potential predators. Cannibalism may also damp density fluctuations, thereby serving as a mechanism of self-regulation. Vermeij and Dudley (1982) reported no evidence of cannibalism in Late Cretaceous assemblages. Recent reanalysis of these assemblages of naticids, however, has revealed evidence of naticid boreholes (Kitchell et al., 1985), suggesting that cannibalism is another behavioral feature of this predator-prey interaction that has been in evidence throughout its evolutionary history. Because prey selection must be assessed in the context of the entire assemblage of available prey and the size structure of both predators and prey, simple indices of intensity of cannibalism are insufficient to reveal whether there has been any directionality in the intensity of cannibalistic behavior.

THE PERSISTENCE OF INEFFICIENT BEHAVIOR

There are not only consequences of inherited structural constraints (e.g., Gould, 1983) but also consequences of inherited behaviors. The naticid predator-prey interaction provides evidence that inefficient behaviors may persist over evolutionarily long periods of time. Naticid predators, when interrupted during drilling, will frequently initiate a new borehole on the same prey item, thereby failing to take advantage of an incipient borehole. Such behavior requires a substantial additional expenditure in drilling time.

FIG. 4-7 Allometry of apertural area (lower slopes) and total cross-sectional area (upper slopes) in the naticid predator *Neverita*. Temporal change is not statistically significant.

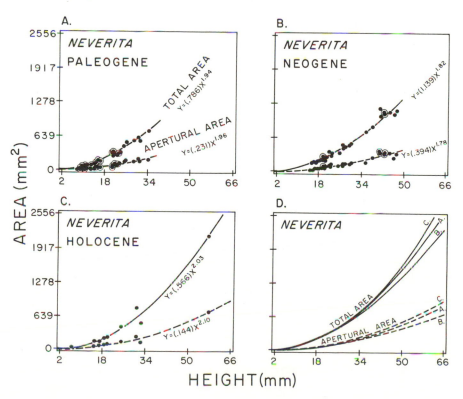

Such inflexible behavior has been documented in the laboratory under controlled conditions on both bivalve and gastropod prey, including the cannibalistic interaction (see Kitchell et al., 1981). We now also have evidence from fossil assemblages that such behavior has been part of the behavioral repertoire early in the evolutionary history of drilling and has continued to persist after tens of millions of years of drilling predation. As emphasized by Lewontin (1983), the evolutionary process is not one that capitalizes on all opportunity.

BEHAVIORAL ENERGETICS: STRATEGIES OF ENERGY ALLOCATION

What are the potential evolutionary consequences of selective predatory behavior? A simulation model of the energetics of both naticid predator and prey was developed to address the question of size change as an evolutionary response to selective predation (DeAngelis et al., 1984). As is apparent from the cost-benefit functions of prey, the predator's net energetic return is a function not only of its behavioral selection of prey size classes and prey species but also of its size. By contrast, two alternate options regarding average prey adult size affect predation rate: increasing growth rate may increase the probability of prey reaching the critical escape size, whereas decreasing growth rate may keep prey sizes in the range of higher relative cost-benefit ratios.

The model incorporates an explicit potential for coevolutionary feedback through size effects: prey size has a direct effect on predator energy intake, and predator size has a direct effect on prey reproductive potential (Figure 4-8). We make the assumption that average predator and prey adult sizes can change over evolutionary time in response to selective pressure. There are no genetic mechanisms contained in the model.

Both predator and prey are permitted simultaneously to seek the average adult size that maximizes net energy return for the predator and maximizes reproductive output for the prey, given arbitrary suboptimal starting conditions. The prey is allowed to change the percentage of its energy that it devotes to growth at different ages; energy diverted to growth is not available for reproduction. The following are determined: (1) the combination of age-dependent prey survival in the face of naticid predation that will maximize prey reproductive value, and (2) the maximization of net energy intake by the predator as a result of predator size and selective behavior. The question addressed by simulation, given the above, is whether coevolution in sizes is predicted when prey and predator simultaneously attempt to maximize their opposing interests, and, second, whether constraints will result in stasis.

The first set of simulations allowed the prey to evolve in the absence

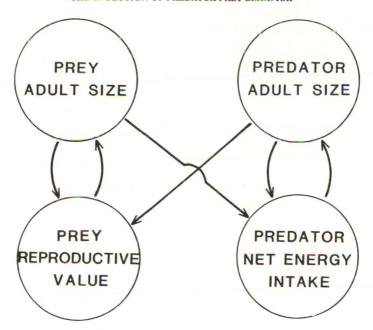

COEVOLUTIONARY FEEDBACK SCHEMATA

FIG. 4-8 Diagram of the positive and negative feedback loops in a model of coevolutionary size change. Predator and prey are connected through the effect of predator size on prey reproductive potential and the effect of prey size on predator energy intake.

of the predator. Regardless of starting conditions, the prey maximized reproductive output by utilizing a strategy of switching from growth to reproduction at a relatively early age (DeAngelis et al., 1984). The second set of simulations included the presence of the predator. The prey's strategy of energy allocation now changed, as did reproductive output. The prey delayed maturation by delaying the age of switching energy allocation from growth to reproduction. The overall result, examined for a variety of initial size conditions, is size change to a stable attracting point (DeAngelis et al., 1984). In particular, the results indicate that in a predator-prey system in which the predator is highly size selective and the probability of successful predation is highly size dependent, there is selection for delayed prey reproduction. Addition of the predator to the prey's "environment" results in the prey delaying the timing of reproduction, thereby facilitating a size response.

Such results also indicate that stasis—in this case, of the parameter of predator and prey size—can result even in predator-prey interactions which act as positive feedback systems. Positive feedback between predator and prey size (Figure 4-8) is constrained at other levels by negative feedback. At large prey sizes, respiratory costs initiate a negative feedback with increasing size whereas at small prey sizes reproduction suffers. Between these constraints, the predator-prey interaction initiates coevolution in size.

The usefulness of these results is that they provide predictions of evolutionary dynamics. Kelley (1984) recently made an explicit test of the predictions of the model presented above and documented an empirical correlation between naticid predation intensity and within-species size increase in several prey species over a three-million-year interval. We present here evidence of an evolutionary trend toward increased size among a suite of naticid predators. The ranges and inferred phylogenetic relationships are given in Marincovich (1977). As illustrated in Figure 4-9, mean size of species of *Polinices, Natica,* and *Neverita* has increased over Cenozoic time. Increase in size within lineages is a common evolutionary trend (Newell, 1949; Gould, 1977), however, and need not be a consequence of size-selective predation.

We were next interested in examining the consequences of dependency between evolution of size and prey morphology. Our earlier research (Kitchell et al., 1981; Boggs et al., 1984) showed that both prey and predator size and prey shell thickness are important components of the prey selection process. The shapes of prey cost-benefit functions are a consequence of prey biomass, shell thickness at the drilling site, and prey size. But these latter two components are competing interests: shell mass must be allocated between thickness of the shell and its size. Consequently, there must be a trade-off between allocation of energy to size vs. thickness. These trade-offs comprise the independent variables of the second model in a series (DeAngelis et al., 1985) developed to explore the potential for coevolution in this predator-prey interaction.

Because this is a dynamic model, there is no prior specification of optimal strategies. There are instead now two dimensions of energy allocation open to the prey: (1) the prey may alter the rate of early growth by the timing of the switch from pure growth to a mixture of growth and reproduction, and (2) the prey may change the relationship of size to thickness, thereby altering shell morphology by diverting additional energy into shell thickness. The third independent variable is the coefficient of the behavioral selection of prey by the predator.

The dependent variables remain the same: (1) the prey attempts to maximize reproductive value as a function of both energy allocation and the probability of survival under naticid predation, and (2) the predator attempts to maximize net energy intake which is similarly a function of prey strategies. The models are complex in that they also incorporate the age structure of both prey and predator populations.

What combination of energy allocation strategies should the prey employ in a biotic environment of energy-maximizing predation? The results clearly demonstrate that the system is dynamic. There is more than one solution to an environment of behaviorally selective preda-

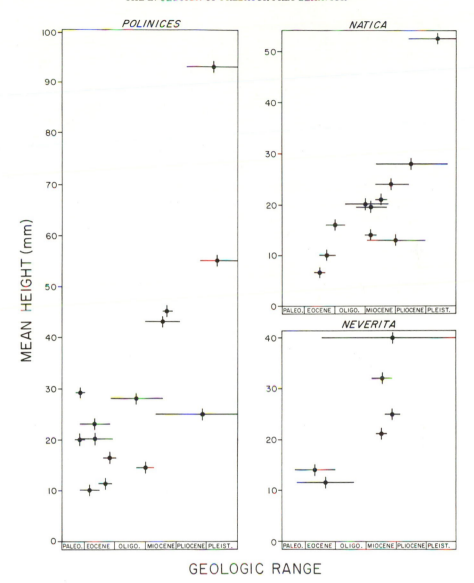

FIG. 4-9 Summary of size data from 31 species of Pacific Coast naticids. Length of horizontal bar denotes known geologic range.

tion. Moreover, the success of any solution not only is constrained by history and architecture but is also context dependent (Kitchell, 1983; DeAngelis et al., 1985).

These results can be displayed as topographic maps of potential prey reproductive output formed by the prey's two intersecting phenotypic strategies. The prey's morphological strategy comprises one axis scaled from 0 to 1 to indicate the proportion of energy allocated to shell length at the expense of shell thickness. The prey's growth-reproduction strategy comprises the other axis. If predation intensity is low, the prey's topographic surface is strongly inclined, indicating lowest repro- ductive output for thick-shelled prey that grow slowly. There is little local relief, indicating that age at first reproduction is less important to reproductive value. If selective predation is intensified to intermediate levels, the phenotypic topography for prey looks quite different. The surface is more wrinkled, indicating sites of local optima, and both overall height and shape of the surface have changed. The most advan- tageous allocation strategies under low predation intensities are now less effective. The relief of the response surface indicates that age at first reproduction is now an important variable (Kitchell, 1983; DeAngelis et al., 1985).

High predation intensity creates even more relief. There is now a relatively large immune area for a specified maximum predator size, and the thin-shelled strategy, regardless of reproductive strategy, has very low potential reproductive output. There is maximum reproduc- tive output for prey with relatively thick strategies and those that repro- duce early do somewhat better.

Overall, these results predict that even smooth continuous gradients, such as those permitted for the prey, may interact with the predator as environment to yield discontinuities. There is also ample potential for nonoptimality: both predator size and predation intensity are variables expected to vary not only on an evolutionary time scale but on a local deme or patch scale.

The significance of these results is that they (1) provide a consistent explanation of modern situations such as that of coexisting species of bivalves with diverse growth and life-history strategies in the face of naticid predation (e.g., Commito, 1982), (2) permit predictions ame- nable to testing in the fossil record (Kitchell et al., 1981, in press; Hoff- man and Martinell, 1984; Kelley, 1984), and (3) stimulate discussion of speciation during periods of intensifying predation (DeAngelis et al., 1985).

CONCLUSIONS

The evolution of predatory behavioral strategies and their evolutionary consequences have been examined:

1. Empirical evidence from experimental work, in situ studies, and fos- sil assemblages refute the hypothesis that naticid predators ran- domly select prey; prey selection is nonrandom and governed by the interaction between energetics and constraints.

2. Despite considerable taxonomic turnover, there is little evidence of evolutionary behavioral change; the same rules of prey selection,

size selection, and site selection pertain to patterns spanning more than 100 million years.

3. Inefficient components of predatory behavior may persist over spans of evolutionary time.

4. Prey selection behavior is more stereotyped than based on individual experience; such stereotypy enhances the usefulness of the temporally integrated data of the fossil record.

5. In addition to between-species selection, size selective behavior has a causal explanation based on energetics and constraints; fossil assemblages document that size selectivity has been a behavioral feature of naticid predation throughout its evolutionary history; there is little empirical evidence of temporal directionality in size selectivity.

6. The predator-prey interaction governed by the behavioral selection of prey is sufficient to generate coevolution in size. The relative fitnesses of both energy allocation and life-history strategies of prey are context-dependent: predatory intensity, maximum predator size, and selective predatory behavior result in dynamic, multiple solutions.

ACKNOWLEDGMENTS

I am grateful to two anonymous reviewers for their helpful comments. This work was supported by NSF grant BSR-8420147.

LITERATURE CITED

Ansell, A. D. 1960. Observations on predation of *Venus striatula* (da Costa) by *Natica alderi* (Forbes). *Proceedings of the Malacological Society of London* 34:157–164.

Berg, C. J., Jr. 1975. A comparison of adaptive strategies of predation among naticid gastropods. *Biological Bulletin* 149:420–421.

Berg, C. J., Jr. 1976. Ontogeny of predatory behavior in marine snails (Prosobranchia: Naticidae). *Nautilus* 90:1–4.

Berg, C. J., Jr. 1978. Development and evolution of behavior in molluscs, with emphasis on changes in stereotypy. In: Burghardt, G. M., and M. Bekoff (eds), The Development of Behavior: Comparative and Evolutionary Aspects. Garland STPM Press, New York, pp. 3–17.

Berg, C. J., Jr., and S. Nishenko. 1975. Stereotypy of predatory boring behavior of Pleistocene naticid gastropods. *Paleobiology* 1:258–260.

Berg, C. J., Jr., and M. E. Porter. 1974. A comparison of predatory behavior among the naticid gastropods *Lunatia heros*, *Lunatia triserata* and *Polinices duplicatus*. *Biological Bulletin* 147:469.

Berry, A. J. 1983. Oxygen consumption and aspects of energetics in a Malaysian population of *Natica maculosa* Lamarck (Gastropoda) feeding on the trochacean gastropod *Umbonium vestiarium* (L.) *Journal of Experimental Marine Biology and Ecology* 66:93–100.

Boggs, C. H., J. A. Rice, J. A. Kitchell, and J. F. Kitchell. 1984. Predation at a snail's pace: What's time to a gastropod? *Oecologia* 62:13–17.

Broom, M. J. 1982. Size-selection, consumption rates and growth of the gastropods *Natica maculosa* Lamarck and *Thais carinifera* (Lamarck) preying on the bivalve *Anadara granosa* (L.). *Journal of Experimental Marine Biology and Ecology* 65:213–233.

Brown, J. L. 1975. The Evolution of Behavior. W. W. Norton, New York.

Commito, J. A. 1982. Effects of *Lunatia heros* predation on the population dynamics of *Mya arenaria* and *Macoma balthica* in Maine, USA. *Marine Biology* 69:187–193.

DeAngelis, D. L., J. A. Kitchell, and W. M. Post. 1985. The influence of naticid predation on evolutionary strategies of bivalve prey: Conclusions from a model. *American Naturalist* 126:817–842.

DeAngelis, D. L., J. A. Kitchell, W. M. Post, and C. C. Travis. 1984. A model of naticid gastropod predator-prey coevolution. In: Levin, S. A., and T. G. Hallam (eds), Mathematical Ecology. Lecture Notes in Biomathematics, No. 54. Springer-Verlag, New York, pp. 120–136.

Edwards, D. C. 1974. Preferred prey of *Polinices duplicatus* in Cape Cod inlets. *Bulletin of the American Malacological Union* 40:17–20.

Edwards, D. C., and J. D. Huebner. 1977. Feeding and growth rates of *Polinices duplicatus* preying on *Mya arenaria* at Barnstable Harbor, Massachusetts. *Ecology* 58:1218–1236.

Franz, D. R. 1977. Size and age-specific predation by *Lunatia heros* (Say, 1822) on the surf clam *Spisula solidissima* (Dillwyn, 1817) off Western Long Island, New York. *The Veliger.* 20:144–150.

Fursich, F. T., and D. Jablonski. 1984. Late Triassic naticid drillholes: Carnivorous gastropods gain a major adaptation but fail to radiate. *Science* 2224:78–80.

Gould, S. J. 1977. Ontogeny and Phylogeny. Harvard University Press, Cambridge, Mass.

Gould, S. J. 1983. Irrelevance, submission and partnership: The changing role of palaeontology in Darwin's three centennials, and a modest proposal for macroevolution. In: Bendall, D. S. (ed), Evolution from Molecules to Men. Cambridge University Press, Cambridge, pp. 347–366.

Griffiths, R. 1981. Predation on the bivalve *Chloromytilis meridionalis* (Kr.) by the gastropod *Natica (Tectonatica) tecta* Anton. *Journal of Molluscan Studies* 47:112–120.

Heiner, R. A. 1983. The origin of predictable behavior. *American Economic Review* 73:560–595.

Hinde, R. A. 1982. Ethology—Its Nature and Relation with Other Sciences. Oxford University Press,

Hoffman, A., and J. Martinell. 1984. Prey selection by naticid gastropods in the Pliocene of Emporda (Northeast Spain). *Neues Jahrbuch fur Geologie und Palaontologie Monatschefte* 1984:393–399.

Hoffman, A., A., Pisera, and M. Ryszkiewicz. 1974. Predation by muricid and naticid gastropods on the Lower Tortonian mollusks from the Korytnica clays. *Acta Geologica Polonica* 24:249–260.

Huebner, J. D., and D. C. Edwards. 1981. Energy budget of the predatory marine gastropod *Polinices duplicatus*. *Marine Biology* 61:221–226.

Kelley, P. H. 1984. Coevolution in a naticid gastropod predatory-prey system: Relation of predation intensity to rates of prey evolution. *Geological Society of America Abstract Program* 16:557.

Kitchell, J. A. 1979. Deep-sea foraging pathways: An analysis of randomness and resource exploitation. *Paleobiology* 5:107–125.

Kitchell, J. A. 1983. An evolutionary model of predator-mediated divergence and coexistence. *Geological Society of America Abstract Program* 15:614.

Kitchell, J. A., C. H. Boggs, J. F. Kitchell, and J. A. Rice. 1981. Prey selection by naticid gastropods: Experimental tests and application to the fossil record. *Paleobiology* 7:533–552.

Kitchell, J. A., C. H. Boggs, J. A. Rice, J. F. Kitchell, A. Hoffmann, and J. Martinell. Anomalies in naticid predatory behavior: A critique and experimental observations. *Malacologia* (in press).

Lewontin, R. C. 1983. Gene, organism and environment. In: Bendall, D. S. (ed), Evolution from Molecules to Men. Cambridge University Press, Cambridge, pp. 273–285.

Marincovich, L., Jr. 1977. Cenozoic Naticidae (Mollusca: Gastropods) of the northeastern Pacific. *Bulletin of American Paleontology* 70:165–494.

Menge, J. L. 1974. Prey selection and foraging period of the predaceous rocky intertidal snail, *Acanthina punctulata*. *Oecologia* 17:293–316.

Morse, D. H. 1980. Behavioral Mechanisms in Ecology. Harvard University Press, Cambridge.

Newell, N. D. 1949. Phyletic size increase, an important trend illustrated in fossil invertebrates. *Evolution* 3:103–124.

Parker, G. A. 1982. Phenotype-limited evolutionarily stable strategies. In: King's College Sociobiology Group (eds), Current Problems in Sociobiology. Cambridge University Press, Cambridge, pp. 173–201.

Schoener, T. W. 1971. Theory of feeding strategies. *Annual Review of Ecology and Systematics* 2:369–404.

Simpson, G. G. 1958. Behavior and evolution. In: Roe, A., and G. G. Simpson (eds), *Behavior and Evolution*. Yale University Press, New Haven, Conn., pp. 507–535.

Sohl, N. F. 1969. The fossil record of shell boring by snails. *American Zoologist* 9:725–734.

Stanton, R. J., Jr., and P. C. Nelson. 1980. Reconstruction of the trophic web in paleontology: Community structure in the Stone City Formation (Middle Eocene, Texas). *Journal of Paleontology* 54:118–135.

Taylor, J. D., N. J. Morris, and C. N. Taylor. 1980. Food specialization and the evolution of predatory prosobranch gastropods. *Palaeontology* 23:375–409.

Thomas, R. D. K. 1976. Gastropod predation on sympatric Neogene species of *Glycymeris* (Bivalvia) from the eastern United States. *Journal of Paleontology* 50:488–499.

Trueman, E. R. 1975. The Locomotion of Soft-Bodied Animals. American Elsevier, New York.

Vermeij, G. J., and E. C. Dudley. 1982. Shell repair and drilling in some gastropods from the Ripley Formation (Upper Cretaceous) of the southeastern U.S.A. *Cretaceous Research* 3:397–403.

Wiltse, W. I. 1978. Effects of predation by *Polinices duplicatus* on community structure. Ph.D. thesis, University of Massachusetts.

II

FIELD AND EXPERIMENTAL APPROACHES TO THE EVOLUTION OF BEHAVIOR

5

Relative Parental Contribution of the Sexes to Their Offspring and the Operation of Sexual Selection

Randy Thornhill

SEXUAL selection is nonrandom differential reproduction of individuals resulting from differential access to mates or the gametes of mates. The term "nonrandom" distinguishes sexual selection from drift, a random process of differential reproduction. The phrase "access to gametes of mates" is added because even though mating success may not vary among individuals of a sex in a nonrandom way, fertilization success may (Parker, 1970; Willson and Burley, 1983; Thornhill and Alcock, 1983; Thornhill, 1983). Sexual selection is a consequence of competition among members of one sex (or their gametes) for members (or gametes) of the opposite sex. The competition among individuals of one sex for the opposite sex may center upon coaxing choosy individuals to mate or to use the competitive sex's gametes, or the competition may simply focus on striving to obtain sexual access to receptive individuals who are willing to mate and reproduce with any conspecific of the opposite sex. The sexual selection which results from the former competition is often referred to as intersexual selection; the latter competition leads to intrasexual selection. The extent of sexual selection acting on a sex is the component of the total variance in reproductive success in that sex which derives from sexual competition for mates or their gametes.

Bateman (1948), Williams (1966), Trivers (1972), and Emlen and Oring (1977) have provided hypotheses for the factors responsible for controlling the operation of sexual selection. Bateman felt that the disparity in gamete sizes of the sexes is the important parameter for understanding the difference in the extent of sexual competition (leading to both intra- and intersexual selection) between the sexes. Williams and Trivers argued that relative parental contribution of the sexes to their offspring is the important factor for understanding sexual selection. Emlen and Oring proposed that the degree of monopolization of the limiting sex (the sex causing sexual competition in the opposite sex) by the limited sex (the sex experiencing sexual competition) is a key variable controlling the operation of sexual selection. They suggested that operational sex ratio (OSR = ratio of sexually active males to receptive females) is an appropriate empirical measure of degree of mate monopolization.

Little research has been directed at critically evaluating these hypotheses. An understanding of the factors controlling sexual selection is important because it is probably a major form of selection in all anisogamous organisms (Darwin, 1874; Bateman, 1948; Trivers, 1972; Alexander and Borgia, 1979; West-Eberhard, 1979; Charnov, 1982; Willson and Burley, 1983; Thornhill and Alcock, 1983), and because the difference in its operation between the sexes may ultimately account for all sexual differences (Trivers, 1972; Power, 1980). I will argue that relative parental contribution of the sexes to offspring is the most important factor accounting for the operation of sexual selection (Williams, 1966; Trivers, 1972). Bateman's view about the disparity in gamete sizes of the sexes does not provide a general theory. Emlen and Oring's idea that degree of mate monopolization is a chief factor controlling the operation of sexual selection is appropriate but in a more restricted sense than they intended. In general, degree of mate monopolization varies as an incidental effect of relative parental contribution of the sexes. My analysis includes results from field experiments with *Panorpa* scorpionflies which show the relationship between factors hypothesized to control the operation of sexual selection.

OPERATION OF SEXUAL SELECTION

Williams, Trivers, and Bateman

Reproductive effort can be defined as the part of an organism's available budget of time, energy, or risk-taking that is directed into reproduction (as opposed to that directed into growth and maintenance) (e.g., Williams, 1966; Hirshfield and Tinkle, 1975). When considering the operation of sexual selection it is useful to distinguish categories of reproductive effort. The two major categories are (1) mating effort, activities, structures, or risk-taking associated with obtaining fertilizations, and (2) nepotistic effort, associated with aiding offspring and other relatives (Low, 1978; Alexander and Borgia, 1979). Trivers (1972) defined parental investment as "any investment [in time, energy, or risk-taking] by the parent in an individual offspring that increases the offspring's chance of surviving (and hence reproductive success) at the cost of the parent's ability to invest in other offspring" (p. 139). Parental effort is a component of nepotistic effort, and an organism's parental effort is the sum of its parental investments because parental investment refers to what a parent does for an individual offspring.

Trivers provided a general theory for relating the operation of sexual selection and the evolution of sexual differences. He concluded, "What governs the operation of sexual selection is the relative parental investment of the sexes in their offspring" (p. 141). By the operation of sexual selection he meant the extent to which one sex (the limited sex) competes for the other (including to be chosen as a mate), measurable by the nonrandom differential reproduction of individuals of the limited sex that is due to differential access to the limiting sex. Trivers's reasoning seems to have been along the following lines. Various biotic and abiotic factors affect the overall reproductive rate of a population (i.e., the number and survival of offspring). Yet parental effort determines the potential reproductive rate of a population by setting an upper limit on offspring number and by reducing the detrimental influences of biotic and abiotic factors that negatively influence the offsprings' chances of survival. Furthermore, since a population is a collection of interbreeding individuals, the parental effort of all the parental individuals of one sex is potentially accessible to each member of the opposite sex. Thus Trivers argued that since parental effort ultimately determines the reproductive rate of a population, parental effort will be the object of all intrasexual competition among members of one sex for the opposite sex, and any sexual disparity in parental effort will cause a level of sexual competition in the sex investing the least parental effort that exactly corresponds to the degree that the opposite sex exceeds it in parental effort. Trivers also argued that it is the differential degree of sexual selection on the sexes that has led to the evolution of sexual differences.

Williams (1966, pp. 183–186) also advanced this view but in a less detailed manner than Trivers. However, Williams did point out that the key factor for understanding the operation of sexual selection is the relative contribution of "materials" and risk-taking of the sexes in providing for the next generation. Thus Trivers and Williams agreed that parental contributions (time, energy, or risk-taking) that affect the reproductive rate of a population (offspring number and survival) are of utmost importance for understanding sexual selection. For example, according to this argument, as the ratio of female-male parental contribution increases there should be a corresponding and direct increase in the extent of sexual selection on males. It should be emphasized that the Williams-Trivers theory addresses the *extent* rather than the *nature* of sexual selection. The evolution of the many ways by which the members of the limited sex compete (e.g., by searching widely for mates, by defending resources critical to their mates, by defending mates directly, or by escalated intrasexual aggression) is a separate issue (for appropriate theory see Trivers, 1972; Bradbury and Vehrencamp, 1977; Emlen and Oring, 1977; Thornhill and Alcock, 1983).

Williams (1966) and Trivers (1972) emphasized that the idea that relative parental contribution controls the operation of sexual selection explains the evolution of sexual differences in reproductive strategy. The male is typically the most sexually competitive sex and the female the most choosy sex. Whereas males typically direct reproductive effort into mating effort, females usually invest more than males into parental effort, and females become the ultimate resource for which the males compete. Female reproductive success is limited by parental effort, and this, coupled with the availability of numerous competing males, provides the basis for the evolution of female choice. Also, they pointed out that sex role reversals occur in species in which males engage in a greater parental effort than females, and there is a reduction in sexual competition when both sexes expend similar parental effort—that is, in the more monogamous species.

Bateman's (1948) theory provided the foundation upon which Trivers (1972) and Williams (1966) built their more general theory. Bateman's principle stems from the fact that sperm are usually very small relative to eggs. For the same allocation of reproductive effort, a male can make vastly more gametes than a female can. Thus, a male's reproductive success is limited by his success at inseminating females and not by ability to invest in gametes, whereas a female's reproductive success is limited by her ability to produce gametes and not by ability to coax a male to inseminate her. This, Bateman argued, accounted for the operation of sexual selection and explained the evolution of eagerness to copulate in males and discrimination and passivity in females.

Williams (1966) and especially Trivers (1972) recognized that to understand what causes sexual selection and to account for its intensity, one must know more than the relative size of male and female gametes. The Williams-Trivers view represents a different theory from Bateman's because parental effort includes all the components of parental contribution to offspring, not just investment in gametes. As a result, hypotheses based on Trivers's theory generate different predictions from those based on Bateman's principle. For instance, if males of a species exhibit greater parental investment per offspring than females, then the males should show less variance in reproduction due to sexual competition than females; this pattern is expected despite the fact that sperm will be smaller than eggs in this species. Although Bateman's principle is consistent with the general difference in male and female strategy, it is an incomplete theory for the operation of sexual selection. Sex role reversals, as well as a reduction in sexual differences under monogamy, are serious exceptions to Bateman's view, but these exceptions provide important support for the Williams-Trivers theory.

Certainty of Parentage and Parental Investment

The arguments of Alexander and Borgia (1979) need to be considered here because they are important for understanding the different operation of sexual selection on the sexes. They emphasize that in addition to the disparity in relative parental effort of the sexes, there are sexual asymmetries in certainty of parentage and the ability to control parental effort allocated to potential offspring or zygotes. In general, males are less certain of parentage than females (i.e., they are less certain of sharing genes with a putative offspring) and have less control than females over any parental effort that is contributed to potential offspring prezygotically or before offspring are born or hatched.

Certainty of parentage and control of investment are related (Alexander and Borgia, 1979). Lower certainty of male than female parentage reduces the likelihood that a male will evolve to invest in his own individual gametes, eggs to be fertilized by them, or zygotes fertilized by them. Regardless of how lower certainty of male parentage occurs, the result is that a male cannot easily control which of his gametes will fertilize eggs. Thus, males are unlikely to benefit from allocating parental effort to individual sperm because of the uncertainty of a male's investment ending up in his zygote. In general, females possess a greater ability to control and benefit reproductively from any resources that they direct to their individual gametes. With internal care of eggs and zygotes by females, males lost control of whether any male-provided resource (e.g., nourishment provided to a female) will end up in an individual egg or zygote; males also lost the ability to control the fate of eggs or zygotes. Females, however, control the fate of individual eggs and zygotes that are internally nourished. A female may absorb eggs, abort zygotes, or digest zygotes internally and eventually redistribute the effort to other eggs and zygotes.

The fitness value associated with allocating reproductive effort into mating effort or parental effort is affected by sexual asymmetries in certainty of parentage and control of investment. In this sense these asymmetries are important for understanding sexual selection. An evolved effect of these asymmetries is a sexual difference in parental effort, which, along with other contributions of the sexes that affect the reproductive rate of the population (see below), controls the operation of sexual selection.

Mating Effort and the Operation of Sexual Selection

Alexander and Borgia (1979) argue that reproductive effort expended by a male prior to formation of a zygote by his mate is mating effort. This includes the effort he expends in the construction and placement of gametes and any resources or aid he may provide to the female prior to zygote formation. Their argument stems from the fact that males have little control over what females do with their own individual eggs. Thus, a male is not likely to be a parental investor before zygote formation because he cannot modify the distribution of his reproductive effort among individual offspring. This ability is necessary for a parental investor, because parental investment refers to a contribution to an individual propagule at the expense of the parent's total parental effort. The same argument can be made for species in which females internally protect and/or nourish eggs after fertilization. It is usually only after fertilized eggs are deposited or young are born that males can have some control of the survivorship of individual progeny and thus become parental investors.

Males of many species supply females with resources prior to, during, or shortly after mating. These resources may be in the form of nuptial gifts of food collected by the male in several species of insects and many birds. In addition, in certain insects, males provide females at mating with nourishment derived from male glandular products, or the male himself may be eaten by the female (for review, see Thornhill and Alcock, 1983).

These resources have been viewed as male parental investment (e.g., Trivers, 1972; Thornhill, 1976; Boggs and Gilbert, 1979; Morris, 1979), because they potentially or actually enhance the survivorship of developing gametes carried by a male's mate. In some cases the nourishment provided by a male goes into eggs he fertilized (see Thornhill and Alcock, 1983). However, in most cases these forms of reproductive effort are best viewed as mating effort because they involve a male's attempt to secure matings, or in some cases they may increase the likelihood that a female will use the sperm of a particular male after it is transferred. That the female uses male-contributed nourishment for her own nutrition or for that of her eggs is inevitable, and not in itself evidence for male parental investment. This is the case regardless of how specialized these male-provided nutrients may be for enhancement of the survival of egg or zygotes.

Gwynne (1984a) has pointed out that Alexander and Borgia's approach is an appropriate way to view partitioning of reproductive effort by the sexes, but that for the purposes of understanding the operation of sexual selection on the sexes, mating effort should be further partitioned into what he calls nonpromiscuous and promiscuous components. The former refers to effort expended by males to obtain fertilizations in which males simultaneously supply material benefits other than sperm to females. These efforts reduce a male's ability to locate other females (i.e., reduce the extent of possible polygyny). Promiscuous mating effort, on the other hand, does not involve these benefits and its expenditure does not reduce the extent of possible polygyny for the male.

Gwynne (1984a) has proposed that since reproductive effort is finite, nonpromiscuous mating effort by reducing the extent of possible polygyny influences the operation of sexual selection in a way similar to parental effort. Also, with both paternal effort and nonpromiscuous mating effort, the benefit males provide to females may be critical to and may limit female reproduction. Thus, to the extent that one sex limits the other's reproduction, selection may favor sexually competitive strategies among members of the limited sex and may favor mate discrimination in the opposite sex whether the limiting factor is eggs, other forms of parental investment, or benefits like food, safety through guarding, and so on. Trivers (1972) also felt that certain types of mating inducements (e.g., nuptial food offerings) by males will influence the operation of sexual selection in the same way as actual parental investment.

Gwynne's argument is correct only in part. The key factor for understanding the relationship between forms of mating effort and the operation of sexual selection is whether the mating effort affects the reproductive rate of the population, and not the extent to which mating effort limits multiple mating. An expenditure of any form of reproductive effort lowers residual reproductive value. Mating investment is any expenditure of time, energy, or risk-taking to obtain a mate that decreases the probability of future matings. Mating effort is the sum of mating investments (Low, 1978; Alexander and Borgia, 1979). Mating effort by one sex that increases offspring number or survivorship limits the reproductive potential of the population and thus will have the same effect as parental effort on the operation of sexual selection, because the expenditure of the mating effort can enhance the reproduction of individuals of the opposite sex and thus will become an object of sexual competition in that sex (Trivers, 1972; Gwynne, 1984a). For example, male mating effort in the forms of food provided to a female or shelter or protection of a female (or protection of a female's offspring if it increases the potential for subsequent matings) may importantly limit female reproduction, even if the mating effort only reduces risk-taking by females in the context of reproduction.

Often mating effort does not limit a population's rate of reproduction. Most organisms are polygynous in the sense that there is greater sexual competition among males for females than vice-versa. Female reproduction in these species is not limited by ability of males to find females, or willingness of males to court or inseminate females. Thus, the male mating effort (time, energy, and risks) directed at finding and inseminating females, or outcompeting other males, does not affect the reproductive rate of the population and will not influence the operation of sexual selection—that is, in this case, the mating effort will not lead to increased sexual competition among females and decreased sexual competition among males. However, if female choice is based on discrimination of sires in terms of their ability to supply genes that influence offspring survival (Trivers, 1972; Zahavi, 1975; Borgia, 1979;

Hamilton and Zuk, 1982; Thornhill and Alcock, 1983), then mating effort that is correlated with presence of high-quality paternal genes may affect the reproductive rate of the population and thereby become a focus of sexual competition among females.

Emlen and Oring

Bradbury and Vehrencamp (1977) and Emlen and Oring (1977) identified the major ecological factors affecting economic monopolizability of mates. In addition, Emlen and Oring (1977) proposed that the degree of mate monopolization controls the operation of sexual selection and provides a predictive theory for understanding the intensity of sexual selection in different mating systems. They suggested that the OSR is a direct measure of ability of members of one sex to monopolize the opposite sex (usually females by males). Under an adult sex ratio of unity, as fewer and fewer males monopolize the females, fewer and fewer females are available, resulting in a corresponding increase in male excess in OSR. For example, according to this approach, when the OSR is biased toward excess males, polygyny is expected, and the intensity of sexual selection on males should be directly related to the degree of the bias. Emlen and Oring (1977) suggested that relative parental investment is only one of several factors influencing OSR and thus only one factor influencing the operation of sexual selection. They pointed out that the usual pattern of relative parental investment of the sexes biases the OSR toward males because males are usually ready to mate and transfer their cheap gametes, but females are less eager and available and they may be tied up for extended periods associated with egg maturation and other forms of parental effort.

However, relative parental investment of the sexes and degree of mate monopolization, under circumstances of both sexes investing (or one sex actually expending parental investment and the other providing a form of mating effort that limits reproduction of the sex actually investing), vary together. And it is relative parental contribution that is the factor controlling the operation of sexual selection, because without parental contribution by one sex there will not be selection for monopolization of that sex by the opposite sex. That is, Williams and Trivers seem to have correctly identified the ultimate factor controlling sexual selection.

First consider Gwynne's (1981, 1984b) excellent work on sexual selection in the katydid *Anabrus simplex*. A major component of male mating effort in *A. simplex* is a large spermatophore (about 25% of male weight), which is produced from food resources acquired by males. The spermatophore material is eaten by a male's mate and it is then used for egg production. Two populations of *A. simplex* were studied, one under high density and the other under low density. Under high density, competition for food apparently makes it more difficult for males to acquire sufficient food for spermatophore production. Males become sexually receptive and willing to mate only when they are ready to transfer a full-sized spermatophore. Few sexually active males are available at any one time under high density. Under low density, most males are continuously sexually active. Thus OSR is biased toward excess females under high density and toward excess males under low density. Under low density males are the competitive sex and females the choosy sex, but under high density the sex roles are reversed. In accordance with the sex role reversal and female excess in the OSR under high density, variance in female mating success is greater, revealing more sexual selection on females in the high-density situation.

Gwynne's (1981, 1984b) estimate of relative parental contribution of the sexes (ratio of average spermatophore weight to average weight of eggs in a clutch) was not different between populations. He felt that parental investment theory would predict a higher ratio under high density. That is, relatively more male contribution than female contribution should be correlated with stronger sexual selection on females and a sex role reversal. Gwynne concluded that the results support Emlen and Oring's (1977) view that degree of mate monopolization is the factor controlling the operation of sexual selection. In *Anabrus*, OSR is a measure of degree of monopolization of the limiting sex by the limited sex. Under high density, as fewer and fewer females obtain the small number of sexually active males available at any one time, females become increasingly overrepresented in the OSR. Under low density, mating by males temporarily reduces the already small number of receptive females available at any one time, and differential mating success among males leads to an excess of males in the OSR that corresponds with the degree of the differential in male success.

OSR change in this katydid system is an epiphenomenon of change in relative parental contribution of the sexes in offspring. Under low density, the high availability of food in the form of plant material apparently leads to many males being sexually ready at any one time, and to a reduced dependence of females on male-provided food in the form of spermatophores. In this situation males control less of what is needed for female reproduction. Under high density, male reproduction is limited by food resource and the ability to convert food to spermatophores, whereas female reproduction is limited by male contribution in the form of spermatophores. That is, male contribution is more important to female fitness under high density, because there are few sexually ready males and many sexually receptive females.

However, it is not spermatophore weight or a ratio of weight of male spermatophore to clutch weight that is important in understanding the role of relative parental contribution in controlling sexual selection and the expression of sexual differences in *A. simplex*. Instead it is the value of a sex's contribution to offspring production of the opposite sex that corresponds with Williams's (1966) and Trivers's (1972) concept of parental contribution (also see Marshall, 1982). In this sense, male *A. simplex* are investing less in offspring production under low density than under high density, and females are investing relatively more under low density than under high density.

In other systems in which males provide benefits important to females, mate monopolization, relative parental contribution of the sexes, and the intensity of sexual selection on males change together. For example, consider a resource-defense polygynous mating system in which males compete for and display a food resource to females that enhances female reproductive success. Females can obtain food on their own without male assistance, but searching for food is risky. If more of the food resource becomes available, more males may obtain it and thereby gain access to females, reducing the intensity of sexual selection on males. As more males contribute the resource (and mate monopolization by males declines), the total contribution from the male sex increases. With greater male contribution, females are required to secure less food on their own and take fewer risks in order to achieve reproductive success—that is, females invest less.

In species in which males provide material benefits that increase the number or survival of offspring, a decrease in the proportion of males providing benefits (number males investing/total number sexually active males)—that is, less investment by males—results in corresponding increases in (1) monopolization of females by males and (2) extent of sexual selection on males. Male-provided material benefits can be subtle. In arthropods, sperm and especially extra-seminal material in the ejaculate are often of nutritional benefit to females and sometimes serve as the basis of female-female sexual competition (for review, see Thornhill and Alcock, 1983). Subtle benefits also include a male protecting his mate from harassment by other males, as in dung flies (Parker, 1978; Borgia, 1979, 1982). In dung flies harassment surrounding mating and egg laying can lead directly to the female's injury or to a delay in time to oviposition, which may increase probability of death before oviposition. These costs of harassment are components of female parental effort. The more risks a female incurs in order to successfully deposit a batch of eggs, the more parental effort she expends. As more male dung flies protect females, the total contribution by males increases (and there are corresponding decreases in mate monopolization by males and variation in mating success of males), and females are increasingly liberated from the risk-taking component of their parental contribution—that is, females invest less. Thus, there is reason to believe that species in which males provide material benefits are numerous and that in such species monopolization of the limiting sex by the opposite sex is an incidental effect of the relative parental contribution of the sexes to offspring.

Wade (1979) and Wade and Arnold (1980) have shown that the variance in mating success divided by the square of the mean mating success is an appropriate statistic describing the intensity of sexual selection. Wade and Arnold's statistic provides a useful empirical measure of the opportunity for evolution by sexual selection if heritable fitness is present. Wade suggested that relative parental contribution of the sexes in offspring is not the factor controlling the operation of sexual selection because it does not enter into the derivation of the statistical measure of the intensity of, or opportunity for, sexual selection. But this statistic is a mathematical description of the intensity of sexual selection, not an evolutionary theory for the biological factors controlling the operation of sexual selection. Regardless of the factor controlling sexual selection, Wade and Arnold's statistic will estimate the resulting intensity of the selection.

In many species, the male's sole contribution to his offspring is the genetic makeup of his sperm. When males contribute only genes, when sperm are abundant, and when sires do not differ in their effect on offspring survival, males provide no contribution that affects the reproductive rate of the population; all parental contribution is expended by females. However, it seems that the extent of sexual selection on males varies under circumstances of complete parental effort by females. For example, Woodward (in manuscript) has shown in work with a species of spadefoot toad that as OSR becomes increasingly biased toward excess males across populations, intensity of sexual selection (measured as $s^2/\overline{x}^2$ mating success; Wade and Arnold, 1980) increases. Woodward feels that the major contribution of male spadefoot toads is sperm and there is little opportunity for males to provide material benefits. The proportion of males contributing sperm (number males mating/number males sexually active) decreases as degree of male excess in the OSR increases, but since a male's only contribution is sperm, the total male contribution is not influenced by OSR change (i.e., every offspring receives one-half of its genes from its father). Regardless of the number of males contributing sperm, the total male contribution when only male genes are involved will not vary as long as female reproduction is not limited by ability to obtain male gametes (or male gametes of high genetic value for offspring survival).

Thus, when there is no male contribution other than sperm, and when sperm are abundant and do not vary significantly in genetic quality relating to offspring survival, degree of mate monopolization by males (and its empirical measure, OSR) determines the extent of sexual selection on males. In this case, relative parental contribution of the sexes controls the operation of sexual selection in terms of the general pattern of strong sexual competition among males for females. Females expend all the parental effort, and the more a male can secure of the total parental effort expended by all females the higher his fitness. But the variation in the extent of sexual selection from population to population and within a population through time will not be caused by variation in parental contribution of the sexes.

NATURAL HISTORY AND MATING SYSTEM
OF PANORPA

In the remainder of this chapter I expand and refine my initial work (Thornhill, 1981) on factors causing variation in the intensity of sexual selection in *Panorpa* scorpionflies. These insects provide an ideal system for experimental analysis of factors controlling the operation of sexual selection. First I discuss the natural history of *Panorpa* and findings from my previous research, and then I discuss results from recent experiments.

Panorpa is the largest genus of North American Mecoptera, with 47 described species (Byers and Thornhill, 1983). *Panorpa* adults are medium-sized insects that live among the herbs of moist forests. Adults of several species of *Panorpa* often occur together in large numbers. They are scavengers, feeding on dead arthropods that they find among the herbs or entrapped in spider webs. When disturbed, they fly only a few feet before coming to rest on the herbs.

Adult *Panorpa* are nocturnal and crepuscular in their mating activities (Byers and Thornhill, 1983). A resource-defense polygynous mating system is exhibited by *Panorpa* species and is characterized by intense sexual competition among males for females, and probably much greater variance in male than in female reproductive success. Male *Panorpa* exhibit three alternative mating behaviors. Individual males are capable of and often employ all three alternatives (Thornhill, 1979a, 1980a, 1981, 1984).

Two alternative mating efforts employed by males involve nuptial feeding, that is, the male presents a food item to the female during courtship and the female feeds on it throughout copulation. In one case a male feeds a female a salivary mass that he secretes. Males must feed in order to produce saliva. After saliva secretion, males stand near the saliva mass and disperse sex pheromone. The pheromone attracts conspecific females at distances up to 8 m (Thornhill, 1979b). Copulation may last a few hours in some *Panorpa* species. Alternatively, a male may feed a female a dead arthropod during copulation. In this case, a male locates a dead arthropod, feeds on it briefly, and then releases sex attractant. During both alternatives, males display with wing movements and abdominal vibrations to females attracted to the pheromone. Also, males defend nuptial offerings of both types from other males that attempt to usurp them through aggression.

The third alternative employed by males is forced copulation. Forced copulation involves a male without a nuptial offering rushing toward a passing female and then grabbing her with his large genital claspers. A male forcing copulation does not release pheromone.

The behavior of females toward males with and without a nuptial offering is distinctly different (Thornhill, 1979a, 1980a, 1984). Females flee from males that approach them without a nuptial offering; however, females approach males with nuptial offerings. Furthermore, females struggle to escape from the grasp of force-copulating males, but females do not resist copulation with resource-providing males. Females manage to escape in 85% of forced copulation attempts. Forced copulation results in insemination 50% of the time, but copulations involving a resource always result in insemination. This difference is not due to reduced sperm levels in force copulators. Evidence indicates that females may, in some way, prevent insemination during some forced copulations (Thornhill, 1984).

The mating system of *Panorpa* has evolved around limited food in the form of dead arthropods. In studies of resource-defense mating systems it is often assumed that the resource contested by males varies in quality and limits the extent of both male and female reproduction. My work on *Panorpa* has avoided this assumption by detailed and long-term study of the important resource and the extent of competition for it. Competition for food in adult *Panorpa* has both intraspecific and interspecific components that often strongly influence fitness of individuals: weight (and hence female fecundity and a male's ability to produce saliva) and survivorship (Thornhill, 1980b, 1981).

Females prefer to mate with males in relation to the following sequence of alternatives: male with large dead arthropod, more than male with small dead arthropod, more than male with saliva. Females actively attempt to avoid force-copulating males under all experimental conditions of food levels and male density I have examined (Thornhill, 1979a, 1980a, 1981, 1984). This pattern of choice by female *Panorpa* is adaptive. First, female preference of mates and female fecundity are positively related (Thornhill, 1984). Females mating with arthropod-providing males lay significantly more eggs per unit time than females mating with saliva-providing males, and females experiencing forced copulation lay very few eggs. Also, females mating with arthropod-providing males have the same fecundity as females that experience super-abundant and varied food. Secondly, it appears that females preferring males with superior nuptial offerings may be required to feed less on their own in order to experience high fecundity and thereby reduce risks associated with foraging. A major group of predators of *Panorpa* is web-building spiders (Thornhill, 1975, 1978, 1981). *Panorpa* sometimes enter spider webs and feed on the arthropods therein. The amount of feeding in webs by *Panorpa* is indirectly related to the amount of food obtained outside webs (Thornhill, 1981).

ALTERNATIVE MALE TACTICS

A field experiment with *P. latipennis* was conducted to determine factors influencing the use of alternative behaviors by males and the relative success of the alternatives. A discussion of the results and design of the experiment provides important background material for an understanding of my study of factors influencing the operation of sexual selection on males. The experiment involved manipulating the abundance of dead arthropods (fresh dead house crickets) available to male *P. latipennis* for use as nuptial gifts. *Panorpa* males cannot move dead arthropods around in the habitat and they attempt to defend a dead arthropod wherever they find it. Experiments were in 3 ft x 3 ft x 3 ft screen enclosures. Only the bottoms of the enclosures were open, allowing their placement among the herbs in *Panorpa* habitat on the Edwin S. George Reserve (University of Michigan) in southeastern Michigan. I used dead cricket abundances of two, four, and six large (about 440 mm) crickets per enclosure. Seven enclosures per cricket abundance were used. The crickets were taped to herbs inside enclosures. I removed old crickets and added fresh crickets at approximately two-day intervals. I weighed, measured (with Vernier calipers) the right forewing length, and individually marked (with model airplane paint) males before adding them to each enclosure. Males and females used in experiments were collected from natural populations. Body size and weight are highly correlated for both sexes. Ten males and ten females were added to each enclosure. Densities similar to this are common in nature. The ten males in each enclosure consisted of three large (55–63 mg), four medium-sized (42–53 mg), and three small (33–41 mg) males. This roughly represents the proportion of males of each size category in natural populations early in the species' adult emergence period. I periodically observed the copulatory behavior of males in each enclosure from approximately sunset to sunrise for about one week, using a Noctron Night Viewer and a flashlight for observations at night. One week is about one-half the mean lifetime of an individual *Panorpa*. Upon the death of a male or female I added a replacement individual to the enclosure. When a male died, I replaced him with a marked individual of the same size class. (Details of techniques for observing individuals in enclosures are discussed in Thornhill, 1981.) The distributions of mating success data among enclosures of each resource level do not differ significantly and thus were combined for analysis. Male body size is positively related to mating success under all three cricket densities (Table 5-1). This was predicted on the basis of my finding that the advantage in male-male competition for dead arthropods in *Panorpa* is strongly positively related to body size (Thornhill, 1979a, 1981). A Kruskal-Wallis test and a Newman-Keuls analysis were used to compare mating success of the three sizes of males in enclosures containing each cricket abundance: *2 crickets:* q (large vs. medium) = 7.9. $p < 0.001$; q (medium vs. small) = 3.0, $0.05 < p < 0.1$. *4 crickets:* q (large vs. medium) = 9.1, $p < 0.001$; q (medium vs. small) = 5.8, $p. < 0.01$. *6 crickets:* q (large vs. medium) = 6.1, $p < 0.01$; q (medium vs. small) = 5.2, $p < 0.01$. The data from two- and four-cricket abundances were collected in 1980; those from six-cricket

TABLE 5-1   Male size in relation to mating success in *P. latipennis* under three resource (dead cricket) abundances. First number = 2 crickets/enclosure (*N* = 201 matings). Number in ( ) = 4 crickets/enclosure (*N* = 170 matings). Number in [ ] = 6 crickets/enclosure (*N* = 150 matings).

Male size	No. males	$\bar{x}$ matings	% of total matings
Large	21 (21) [21]	5.8 (3.9) [3.1]	60 (48) [43]
Medium	28 (28) [28]	1.9 (2.3) [2.1]	27 (37) [40]
Small	21 (21) [21]	1.2 (1.2) [1.2]	13 (15) [17]

abundance were collected in 1981. Thus, it may be inappropriate to compare the results from the two- and four-cricket densities with those from the six-cricket density. However, the results obtained in 1981 are precisely in the direction expected, and other resource manipulations conducted in 1981 (see below) reveal the same relationship between cricket abundance and mating success obtained in 1980.

Males of the three sizes employ different repertoires of alternative behaviors, and the proportion of time a male was observed courting using a given alternative reflects his body size and the extent of competition for dead crickets (Table 5-2). For example, under all three

TABLE 5-2   Male mating success and alternative behaviors used by males in *P. latipennis* in relation to body size under three resource (dead cricket) abundances. First number = 2 crickets/enclosure (*N* = 201 matings). Number in ( ) = 4 crickets/enclosure (*N* = 170 matings). Number in [ ] = 6 crickets/enclosure (*N* = 150 matings).

Male size	% mating using		
	Cricket	Saliva	Force
Large	71 (89) [96]	27 (11) [4]	3 (0) [0]
Medium	22 (35) [46]	60 (57) [52]	18 (8) [2]
Small	4 (4) [12]	54 (58) [56]	42 (38) [32]
All 3 sizes	49 (57) [61]	39 (35) [33]	12 (8) [6]

cricket densities large males use dead arthropods more often than saliva to obtain matings, and large males rarely use force. Use of crickets by large males during mating increases with cricket density, and the percentage of matings involving saliva decreases in a corresponding fashion. Similar trends are seen with the other size categories of males. Furthermore, data in the last row of Table 5-2 reveal that providing a cricket results in higher mating success under all three resource levels, and the next most successful alternative is use of saliva. The reproductive success of forced copulators is even lower than depicted in Table

5-2 because 50% of forced matings do not result in insemination (Thornhill, 1984).

The results from the field experiment show that the pattern of female preference for males employing the three mating alternatives also depicts relative male reproductive success (as measured by copulatory success) associated with the alternatives. The field experiment as well as laboratory experiments (Thornhill, 1979a, 1980a, 1981, 1984) reveal that male *Panorpa* prefer to use dead arthropod presentation more than saliva presentation, and that males resort to forced copulation only when the other two alternatives cannot be adopted because of resource scarcity. This scarcity may stem from low absolute food abundance or from increased male-male competition, or both. The field and laboratory work also indicate that male body size and weight determine a male's success in male-male competition (direct aggression) for resources, and the larger the male the more he employs alternatives resulting in greater relative mating success. (Scorpionflies do not grow in size after they molt to adulthood.)

RESOURCE ABUNDANCE
AND SEXUAL SELECTION

The field experiment discussed above was also conducted to determine the relationship between the intensity of sexual selection on males and resource levels in a resource-defense polygynous system. The prediction tested was that the intensity of sexual selection will decline as dead arthropod abundance increases. This prediction stems directly from the notion that relative parental contribution of the sexes controls the operation of sexual selection in systems in which both sexes make a parental contribution. In this case, as the abundance of dead arthropods is increased, the proportion of males contributing resource and the total contribution by males also increase. Note that ability of males to monopolize mating opportunities also decreases as dead arthropods become more available.

A summary of the data on variance in male mating success obtained from the enclosure experiment is contained in Table 5-3. The results were treated two ways. First, I calculated mean and variance in mating success on the basis of total sample sizes of matings (samples in Table 5-1; see Thornhill, 1981, for the outcome of this approach for 2-cricket/enclosure and 4-cricket/enclosure resource densities). Second, I drew 140 copulations at random from the data for the three resource levels and based mean and variances on these samples (as in Table 5-3). The second method was used because total number of copulations observed differed among resource levels. Both methods show the same pattern: variance in male mating success declines as resources are added (Thornhill, 1981). In Table 5-3, variance alone describes the

TABLE 5-3 The extent of sexual selection in male *P. latipennis* (s^2 in matings achieved) in relation to different levels of resources (F_{max} — test: $F_{max} = 2.53$, $p < 0.01$).

| Resource level | No. males | Matings | | | Fit to Poisson |
		No.	$\bar{x}$	s^2	
Enclosures with 2 crickets	70	140	2.0	6.4	No; $\chi^2 =$ 37.1 $p < 0.005$
Enclosures with 4 crickets	70	140	2.0	4.1	No; $\chi^2 =$ 22.2 $p < 0.005$
Enclosures with 6 crickets	70	140	2.0	2.5	Yes; $\chi^2 = 6.0$ $p = 0.10$

extent of sexual competition among males because mean mating success is the same across dead-cricket densities. Of course, Wade and Arnold's (1980) recommended measure of actual intensity of sexual selection on males ($s^2/\bar{x}^2$) also declines in Table 5-3 as cricket density increases.

An F_{max} test indicates a significant decline in variance as resources become more available (Table 5-3). F tests are not robust tests for equality of variances (e.g., Zar, 1974; van Valen, 1978). Levene's test is better (see van Valen, 1978) and when applied to the results summarized in Table 5-3 reveals that all pair-wise comparisons of the three variances are significantly different ($p \leq 0.05$).

I include in Table 5-3 the fit of each data set to the Poisson distribution. As resources increase, the Poisson distribution is approached and met by mating success data from the highest resource density. However, sexual competition is still occurring under this resource level—that is, male mating success is related to body size (Table 5-1). The Poisson distribution assumption of independence of observations probably is violated in this experiment, making the Poisson distribution an inappropriate null hypothesis. A male's success or failure in the future may depend importantly on his success or failure in the past (e.g., see Alexander, 1961).

It is important to remember in studies of intensity of selection that the variance in success among individuals may be generated by random events. If so, the variation is not related to selection but instead describes the opportunity for drift. Given that the Poisson distribution is probably an inappropriate null hypothesis in studies of social behavior, it is imperative to study the cause(s) of the variation in individual performance to determine if the variation is random or nonrandom with regard to phenotypes (also see Fincke, 1982; McCauley, 1983; Payne, 1984).

SEX RATIO AND SEXUAL SELECTION

In a separate field enclosure experiment in 1981, I began an examination of the relationship among OSR, relative parental contribution of the sexes, and the intensity of sexual selection on males in *P. latipennis*. The prediction tested was that as adult sex ratio becomes more female-biased (i.e., as male monopolization of females declines and a greater proportion of males contribute resource), the intensity of sexual selection on males will decline. Seven replicates of three different adult sex ratios were used: 15 males/5 females, 10 males/10 females, 5 males/15 females. I kept dead-arthropod density (four large crickets per enclosure) and adult *Panorpa* density constant across enclosures. Male size distributions were as similar as possible to those used in the previous experiment, given that male number varied across sex ratio treatments (sex ratio/number of large, medium, and small males used for each sex ratio: 3/4, 7, 4; 1/3, 4, 3; 0.33/2, 2, 1). The only variable manipulated was adult sex ratio; as a result, food levels per male also varied as in the previous experiment. I chose this experimental design rather than equal resource per male across enclosures in order to determine if sex ratio changes would affect the pattern obtained in the previous experiment. The pattern is the same (Table 5-4). As the number of females increases relative to the number of males, or as the number of males per cricket declines and a greater proportion of males invest crickets, the extent of sexual selection on males ($s^2/\overline{x}^2$ mating success) decreases. The data in the second row (sex ratio 1:1; 10 males/10 females) are from the previous experiment. As with the mating success data from a 1:1 sex ratio, the mating success data from each of the other two experimental sex ratios did not differ significantly and were lumped for analysis.

This field experiment employed females of variable reproductive state. Females and males were collected from natural populations and placed in enclosures. In 1983, I investigated the influence of OSR per

TABLE 5-4 The extent of sexual selection in male *P. latipennis* in relation to adult sex ratio.

Sex ratio	No. males	Matings			
		No.	$\overline{x}$	s^2	$s^2/\overline{x}^2$
15 males:5 females in each of 7 enclosures	105	141	1.3	4.5	2.7
10 males:10 females in each of 7 enclosures	70	140	2.0	4.1	1.0
5 males:15 females in each of 7 enclosures	35	141	4.0	4.4	0.3

se (ratio of sexually active males to *receptive* females in the population) on intensity of sexual selection on males in *P. latipennis*. Female sexual receptivity is easily determined in *Panorpa*. Field-collected females are placed in a cage that contains a smaller caged male releasing phero-mone. Only receptive females respond by aggregating on and near the caged male (Thornhill, 1979b). Males are more or less continually sex-ually receptive (e.g., Thornhill, 1981). Males were fed (same amount of food per male: two large crickets) for 24 hours prior to each test so that they were all capable of secreting several salivary masses as nuptial gifts (for techniques, see Thornhill, 1981). Males were introduced into enclosures in late afternoon, and when they had secreted saliva and began pheromone liberation at dusk, receptive females were intro-duced. The use of saliva providing males rather than arthropod defend-ers eliminates the difficulty of keeping resource levels per male con-stant across the OSR manipulation (see above). Overall density of *Panorpa* was kept constant across enclosures. All males were distinctly marked and measured. Number of replicates, male size distributions, and sex ratio were as in the previous experiment. Observations began at dusk on June 15 and continued until mating activities ceased the following dawn (experimental period about 12 hours). Enclosures were checked at hourly intervals and mating and nonmating individuals recorded. The opportunity for female choice exists in this experiment even though all males can produce saliva offerings because male size and size of saliva mass are positively correlated (Thornhill, 1981).

The majority of females were observed mating only once. All of the 35 females under the sex ratio of 3 (15 males/5 females) mated, and 8 mated more than once. Only one of these females mated three times. Only 2 of the 70 females under the 1:1 sex ratio (10 males/10 females) mated twice, and one female did not mate. Under the sex ratio of 0.33 (5 males/15 females), one female mated two times and one did not mate. The infrequent mating of female *Panorpa* is due to the onset of a one- to two-day period of sexual nonreceptivity after matings in which males provide a resource, either a dead arthropod or saliva (Thornhill, 1984).

As expected, the intensity of sexual selection ($s^2/\bar{x}^2$ mating success) on males declines as OSR becomes less biased toward excess males (Table 5-5). Female *Panorpa* are expected to accept more readily infe-rior (i.e., smaller) males as mate choice options are reduced (i.e., when fewer males are available), especially since there is probably a selective disadvantage associated with long delays in mating by females in sea-sonal breeders like *Panorpa*. This was examined by comparing the mat-ing success of males of different sizes under the three sex ratios (Tables 5-6, 5-7, and 5-8). Under all three sex ratios large males had the highest mean mating success, medium-sized males next, and small males the least, and intensity of sexual selection on each size category of males declines as sex ratio (male/female) declines. Also, in general under all

TABLE 5-5   The extent of sexual selection in male *P. latipennis* in relation to operational sex ratio.

Sex ratio	No. males	Male matings						% males unmated
		No.	Range	$\bar{x}$	s^2	$s^2/\bar{x}^2$		
15 males:5 females in each of 7 enclosures	105	45	0–4	0.43	0.64	3.46		67.62
10 males:10 females in each of 7 enclosures	70	72	0–4	1.03	1.15	1.08		41.43
5 males:15 females in each of 7 enclosures	35	105	0–5	3.00	1.24	0.14		2.86

TABLE 5-6   Male mating success in relation to male body size under an operational sex ratio of 15 males:5 females in *P. latipennis*.

	Males			Matings					
Male size	No.	Proportion	% Unmated	No.	Range	% Total	$\bar{x}$	s^2	$s^2/\bar{x}^2$
Large	28	0.27	0.95	38	0–4	84.44	1.36	0.61	0.33
Medium	49	0.47	87.76	6	0–1	13.33	0.12	0.11	7.86
Small	28	0.27	96.42	1	0–1	2.22	0.04	0.04	25.00

TABLE 5-7   Male mating success in relation to male body size under an operational sex ratio of 10 males:10 females in *P. latipennis*.

	Males			Matings					
Male size	No.	Proportion	% Unmated	No.	Range	% Total	$\bar{x}$	s^2	$s^2/\bar{x}^2$
Large	21	0.30	0.0	51	1–4	70.83	2.43	0.66	0.11
Medium	28	0.40	46.43	16	0–2	22.22	0.57	0.33	1.03
Small	21	0.30	76.19	5	0–1	6.94	0.24	0.19	3.17

TABLE 5-8   Male mating success in relation to male body size under an operational sex ratio of 5 males:15 females in *P. latipennis*.

	Males			Matings					
Male size	No.	Proportion	% Unmated	No.	Range	% Total	$\bar{x}$	s^2	$s^2/\bar{x}^2$
Large	14	0.40	0.0	54	2–5	51.43	3.86	0.747	0.050
Medium	14	0.40	0.0	40	2–4	38.10	2.86	0.286	0.035
Small	7	0.20	14.29	11	0–2	10.48	1.57	0.619	0.252

TABLE 5-9 Relationship among variation in OSR, mate monopolization, intensity of sexual selection, and relative parental contribution of the sexes in an experiment with *P. latipennis*.

OSR	Mate monopolization (% males unmated)	Intensity of sexual selection on males	Percentage of males providing saliva	Percentage of females investing	Reproductive potential (no. eggs) of females*	Contribution of males to reproduction of population
3	67.6	3.5	32.4	100	1750	1750/105 = 16.7
1	41.4	1.1	58.6	100	3500	3500/70 = 50
0.33	2.9	0.1	97.1	100	5250	5250/35 = 150

*Assumes each female will lay 50 eggs following mating with a saliva-providing male. This is generally the case (Thornhill, 1984).

sex ratios sexual selection is more intense on small males than on medium-sized males, and large males experience the least sexual selection.

In this experiment variation in the OSR coincides with changes in relative parental contribution of the sexes as well as the degree of female monopolization by males (Table 5-9). Mate monopolization by males is measured as percent of males unmated. All males attempted to mate, but as OSR became more male biased toward excess males, more males were excluded from mating. More males provided food to females as OSR favored females, but, regardless of OSR, essentially all females (only two females did not mate) would go on to contribute to the reproductive rate of the experimental population (i.e., lay a batch of eggs). Females lay eggs during the one to two days of nonreceptivity following mating. Egg batches vary with female size but average about 50 eggs each (Thornhill, 1984). Thus, the reproductive potential of the females under each sex ratio can be calculated by multiplying the number of females by the average number of eggs per batch. Finally, the contribution of males to the reproductive rate of the population is related to the female contribution just mentioned divided by the number of males potentially contributing resources. This ratio increases as OSR becomes less biased toward excess males; that is, the average male parental contribution increases as OSR decreases.

Overall, the experiments involving *Panorpa* show the relationship among factors felt to be important in controlling the operation of sexual selection. It is clear that Bateman's (1948) theory is inappropriate because gamete sizes of the sexes remain the same in all experiments and treatments yet the intensity of sexual selection changes. The experiment dealing with manipulation of OSR per se examines in the most detail the relationship among changes in relative parental contribution of the sexes to their offspring, OSR, degree of female monopolization by males, and intensity of sexual selection on males. All of these factors vary together. But it is relative parental contribution of the sexes that is the ultimate causal variable changing the intensity of sexual selection on males, because only parental contribution affects the number and survival of offspring. In theory, as males contribute more parentally in the OSR experiment the extent of sexual selection on females should increase, but I have not arrived at a satisfactory measure of the degree of sexual selection on female *Panorpa*. A relevant empirical measure of sexual selection on females would be the variation in female reproductive success due to sexual competition for mates. However, the fact that female *Panorpa* seem to relax mate-choice standards when male parental contribution is relatively high (and OSR is biased toward excess females) implies an increase in sexual competition among females in relation to increased paternal contribution.

ACKNOWLEDGMENTS

I thank Gerry Dizinno, Gary Dodson, Darryl Gwynne, Jennifer Kitchell, Bill Kuipers, Larry Marshall, Matt Nitecki, Nancy Thornhill, Bruce Woodward, and two anonymous reviewers for their helpful comments on the manuscript. Stan Braude, Mike Brown, and Nancy Thornhill provided expert research assistance. Bill Dawson and Ron Nussbaum provided housing and research facilities at the Edwin S. George Reserve, University of Michigan. My research on sexual selection has been generously supported by the National Science Foundation (BNS-7912208, DEB-7910193, BSR-8219810, BSR-8305774).

LITERATURE CITED

Alexander, R. D. 1961. Aggressiveness, territoriality and sexual behavior in field crickets (Orthoptera: Gryllidae). *Behaviour* 17:130–223.

Alexander, R. D., and G. Borgia. 1979. On the origin and basis of the male-female phenomenon. In: Blum, M. S., and N. A. Blum (eds), Sexual Selection and Reproductive Competition in Insects. Academic Press, New York, pp. 417–440.

Bateman, A. J. 1948. Intra-sexual selection in *Drosophila*. *Heredity* 2:349–368.

Boggs, C. L., and L. E. Gilbert. 1979. Male contribution to egg production in butterflies: Evidence for transfer of nutrients at mating. *Science* 206:83–84.

Borgia, G. 1979. Sexual selection and the evolution of mating systems. In: Blum, M. S., and N. A. Blum (eds), Sexual Selection and Reproductive Competition in Insects. Academic Press, New York, pp. 19–80.

Borgia, G. 1982. Experimental changes in resource structure and male density: Size related differences in mating success among male *Scatophaga stercoraria*. *Evolution* 36:307–315.

Bradbury, J. W., and S. L. Vehrencamp. 1977. Social organization and foraging in emballonurid bats. III. Mating systems. *Behavioral Ecology and Sociobiology* 2:1–17.

Byers, G. W., and R. Thornhill. 1983. Biology of the Mecoptera. *Annual Review of Entomology* 28:203–228.

Charnov, R. 1982. Sex Allocation. Princeton University Press, Princeton, N.J.

Darwin, C. 1874. The Descent of Man and Selection in Relation to Sex., 2nd edition. A. L. Bunt, New York.

Emlen, S. T., and L. W. Oring. 1977. Ecology, sexual selection and the evolution of mating systems. *Science* 197:215–223.

Fincke, O. M. 1982. Lifetime mating success in a natural population of the damselfly, *Enallagma hageni* (Walsh) (Odonata: Coenagrionidae) *Behavioral Ecology and Sociobiology* 10:293–302.

Gwynne, D. T. 1981. Sexual difference theory: Mormon crickets show role reversal in mate choice. *Science* 213:779–780.

Gwynne, D. T. 1984a. Male mating effort, confidence of paternity and insect sperm competition. In: Smith, R. L. (ed), Sperm Competition and the Evolution of Animal Mating Systems. Academic Press, New York, pp. 117–149.

Gwynne, D. T. 1984b. Sexual selection and sexual differences in Mormon crickets (Orthoptera: Tettigoniidae, *Anubrus simplex*). *Evolution* 38:1011–1022.

Hamilton, W. D., and M. Zuk. 1982. Heritable fitness and bright birds: A role for parasites. *Science* 218:382–387.

Hirschfield, M. F., and D. W. Tinkle. 1975. Natural selection and the evolution of reproductive effort. *Proceedings of the National Academy of Sciences* (USA) 72:2227–2231.

Low, B. S. 1978. Environmental uncertainty and the parental strategies of marsupials and placentals. *American Naturalist* 112:197–213.

McCauley, D. E. 1983. An estimate of the relative opportunities for natural and sexual selection in a population of milkweed beetles. *Evolution* 37:701–707.

Marshall, L. D. 1982. Male nutrient investment in the Lepidoptera: What nutrients should males invest? *American Naturalist* 120:27–35.

Morris, G. K. 1979. Mating systems, paternal investment and aggressive behavior of acoustic Orthoptera. *Florida Entomologist* 62:9–17.

Parker, G. A. 1970. Sperm competition and its evolutionary consequences in the insects. *Biological Review Cambridge Philosophical Society* 45:525–567.

Parker, G. A. 1978. Searching for mates. In: Krebs, J. R., and N. B. Davis (eds), Behavioural Ecology, an Evolutionary Approach. Blackwell, London, pp. 214–244.

Payne, R. B. 1984. Sexual selection, lek and arena behavior, and sexual size dimorphism in birds. *Ornithological Monographs* 33:1–52.

Power, H. 1980. The foraging behavior of mountain bluebirds, with emphasis on sexual differences in foraging. *Ornithological Monographs* 28:1–52.

Thornhill, R. 1975. Scorpionflies as kleptoparasites of web-building spiders. *Nature* 253:709–711.

Thornhill, R. 1976. Sexual selection and paternal investment in insects. *American Naturalist* 110:153–163.

Thornhill, R. 1978. Some arthropod predators and parasites of adult Mecoptera. *Environmental Entomology* 7:714–716.

Thornhill, R. 1979a. Male and female sexual selection and the evolution of mating strategies in insects. In: Blum, M. S., and N. A. Blum (eds), Sexual Selection and Reproductive Competition in Insects. Academic Press, New York, pp. 81–122.

Thornhill, R. 1979b. Male pair-formation pheromones in *Panorpa* scorpionflies (Mecoptera: Panorpidae). *Environmental Entomology* 8:886–889.

Thornhill, R. 1980a. Rape in *Panorpa* scorpionflies and a general rape hypothesis. *Animal Behaviour* 28:52–59.

Thornhill, R. 1980b. Competition and coexistence among *Panorpa* scorpionflies (Mecoptera: Panorpidae). *Ecological Monographs* 50:179–197.

Thornhill, R. 1981. *Panorpa* (Mecoptera: Panorpidae) scorpionflies: Systems for understanding resource-defense polygyny and alternative male reproductive efforts. *Annual Review of Ecology and Systematics* 12:355–386.

Thornhill, R. 1983. Cryptic females choice and its implications in the scorpionfly *Harpobittacus nigriceps. American Naturalist* 122:765–788.

Thornhill, R. 1984. Alternative hypotheses for traits presumed to have evolved in the context of sperm competition. In: Smith, R. L. (ed), Sperm Competition and the Evolution of Animal Mating Systems. Academic Press, New York, pp. 151–178.

Thornhill, R., and J. Alcock. 1983. The Evolution of Insect Mating Systems. Harvard University Press, Cambridge, Mass.

Trivers, R. L. 1972. Parental investment and sexual selection. In: Campbell, B. (ed), Sexual Selection and the Descent of Man, 1871–1971. Aldine, Chicago, pp. 136–179.

Van Valen, L. 1978. The statistics of variation. *Evolutionary Theory* 4:33–43.

Wade, M. J. 1979. Sexual selection and variance in reproductive success. *American Naturalist* 114:742–747.

Wade, M. J., and S. J. Arnold. 1980. The intensity of sexual selection in relation to male sexual behavior, female choice and sperm precedence. *Animal Behaviour* 28:446–461.

West-Eberhard, M. J. 1979. Sexual selection, social competition and evolution. *Proceedings of the Philosophical Society of America* 123:222–234.

Williams, G. C. 1966. Adaptation and Natural Selection. Princeton University Press, Princeton, N.J.

Willson, M. F., and N. Burley. 1983. Mate Choice in Plants: Tactics, Mechanisms, and Consequence. Princeton University Press, Princeton, N.J.

Woodward, B. In manuscript. Intensity of sexual selection in spadefoot toads.

Zahavi, A. 1975. Mate selection—a selection for a handicap. *Journal of Theoretical Biology* 53:205–214.

Zar, J. H. 1974. Biostatistical analysis. Prentice-Hall, Englewood Cliffs, N.J.

6

Demographic Routes to Cooperative Breeding in Some New World Jays

John W. Fitzpatrick and Glen E. Woolfenden

COOPERATIVE breeding—the contribution by non-parental individuals to rearing offspring—is known to characterize hundreds of bird species in a wide variety of families (Brown, 1978; Emlen and Vehrencamp, 1983). In recent years, advances in both theory and field investigations have converged upon a general observation regarding the ecological settings in which the phenomenon arises. Numerous reviews now are available, each one stating the ecological conditions somewhat differently, depending mainly upon the specific case studies most familiar to the reviewer (e.g., Brown, 1978; Gaston, 1978; Emlen, 1978, 1982a, 1982b, 1984; Emlen and Vehrencamp, 1983; Vehrencamp, 1979; Koenig and Pitelka, 1981; Woolfenden and Fitzpatrick, 1984). We summarize the underlying consensus as follows: cooperative breeding among birds arises in ecologically peculiar settings, where free *access to successful breeding* is in some way limited. Access can be restricted through limitation or spatial localization of some critical resource, or through several kinds of effects of extreme climatic variability (Koenig and Pitelka, 1981; Emlen and Vehrencamp, 1983). Most often, but not always, this limitation most strongly affects pre-reproductive individuals (see Reyer, 1980, for a counter-example).

In this chapter we discuss these ecological conditions from a demographic perspective, as they pertain to the Florida Scrub Jay social system in particular and to the corvid genus *Aphelocoma* as a whole. This genus of New World jays provides an especially revealing framework for examining social evolution. It includes closely related populations that exhibit a range of social systems varying from lone-pair territories with early-dispersing offspring to multiple-pair territories in which offspring delay dispersal and even breed within the natal group (Figure 6-1; Verbeek, 1973; Brown, 1974). We bring to this analysis 15 years' data from an intensive, ongoing study of a marked population of Florida Scrub Jays. Our goal is to construct a theoretical framework that predicts the ecological conditions which would favor the various group sizes and social structures found within this instructive phylogenetic assemblage. We hope the approach, if not the conclusions, will be valid in the general case as well.

Our focus in this report is demographic. In our view social behavior is a life history trait, conceptually inseparable from demographic features such as birth rates, death rates, and movement patterns. Interpreting social behavior therefore requires detailed data on these demographic features within real populations. Frequently such data in turn permit realistic modeling of alternative life history strategies or social traits as a way of predicting conditions under which other social systems should evolve.

We address four main points: (1) Where habitat is a limiting resource, the presence or absence of suboptimal habitat, and its suitability for survival and breeding, become critical factors; sharp boundaries between usable and unusable habitat eliminate an important dispersal option as pre-reproductives attempt to become breeders; the usefulness of so-called marginal habitat can be defined and empirically measured in demographic terms. (2) Within usable habitat, production of offspring that reach breeding age can exceed the rate at which vacancies appear in the breeding population; the degree of such crowding can be quantified. (3) Delayed dispersal and cooperative breeding can be favored if early dispersal is sufficiently more costly, or if annual probabilities of obtaining a breeding position can be increased with age. (4) If juvenile mortality is high and variance in lifetime reproductive success of breeders also is high, extended investment by parents in their older non-breeding offspring is favored, theoretically even at the expense of present reproduction.

The existence and number of non-breeding helpers within a population of cooperative breeders are predictable outcomes of the above considerations. Multi-pair territories represent an extreme along a continuum of social stages corresponding to the degree of competition for access to breeding resources within the local population. Territorial budding represents an intermediate condition along this continuum.

STUDY SITE AND METHODS

Since 1969 we have followed from birth to death several hundred Florida Scrub Jays that have lived in a study tract of about 400 ha. The tract is part of the Archbold Biological Station, a 1700 ha field research station devoted to long-term studies of animals and plants native to the unique Florida oak scrub. The station is located within the largest and finest remaining stand of this rare and vanishing habitat within south central Florida.

For over a decade, all but one or two of the 100 to 200 jays residing in the study tract have been individually marked with colored plastic leg bands. Each month since 1971, all members of about 30 jay families have been censused. Reproduction in each of these families is closely monitored every spring, and each new cohort of young jays is marked as nestlings. About 75% of the breeders in our population presently are of known age and known parentage, and this proportion has reached equilibrium. The few jays that immigrate into the tract each year and become established as breeders are captured and marked. We regularly search the extensive area of jay habitat surrounding our tract and record any successful, marked dispersers. Because Florida Scrub Jays are extremely sedentary, such emigrations are rare, and we find most of them. Extensive discussion of our methods and of the study area appeared in a recent monograph (Woolfenden and Fitzpatrick, 1984).

THE GENUS APHELOCOMA

The genus *Aphelocoma* consists of three closely related species of jays (family Corvidae). The breeding systems of various populations of two species, *A. coerulescens*, the Scrub Jay, and *A. ultramarina*, the Mexican Jay, are well studied and show an intriguing array of variation (Figure 6-1). Scrub Jays from mainland western North America breed as unassisted pairs. Most pairs dwell in permanent territories, but those living in regions with severe winters move to lower elevations during the non-breeding season (F. Pitelka, pers. comm.). In all areas in western U.S., offspring disperse permanently a few months after hatching. Probably many individuals breed at age one year, as reported by Ritter (1983). Thus, the widespread western populations of Scrub Jays represent the social system that is most common among birds. The distinct Santa Cruz Island population, *A. c. insularis*, differs from its mainland relatives in that breeding often is delayed. Troops of non-breeding jays, which include individuals several years old as well as many yearlings, roam through *marginal* habitat not contained in the territories of breeders. These non-breeders repeatedly intrude into occupied land, seemingly testing for weaknesses or gaps among the territorial breeders. Some Santa Cruz jays do breed at age one year (Atwood, 1980).

Florida Scrub Jays, *A. c. coerulescens*, represent another stage in the continuum, where both dispersal and breeding are delayed. Florida Scrub Jays do not breed as yearlings (defined as individuals between ages 12 and 22 months), and many do not breed until several years old. Prior to breeding, these jays remain in their natal territories where they assist the breeders, who usually are their parents. Breeders with helpers produce more offspring and show decreased mortality compared to breeders without helpers (Woolfenden and Fitzpatrick, 1984). Non-breeders practice dispersal forays away from their natal territory, and

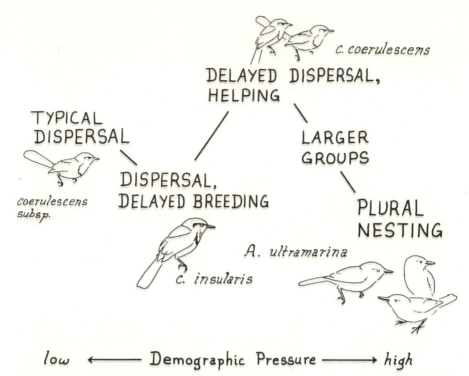

FIG. 6-1 Hypothesized pathways of social evolution in the corvid genus *Aphelocoma*. Each stage pictured is represented by known populations of *C. coerulescens* or *A. ultramarina*. Different stages are hypothesized to reflect different degrees of demographic pressure, in turn caused by different degrees of habitat limitation or crowding.

these absences can last from a few minutes up to several weeks or months, after which the jays return home. Permanent departure from the natal territory typically occurs only when the disperser pairs with another jay and begins to defend its new territory as a breeder. Males and females follow quite different dispersal strategies (Woolfenden and Fitzpatrick, in press), with females dispersing both earlier in their lives and farther from home. Males frequently inherit breeding space by acquiring an immigrant mate and defending a segment of the natal territory—a process we refer to as territorial budding (Woolfenden and Fitzpatrick, 1978).

Mexican Jays differ from Florida Scrub Jays in that territories can contain more than one breeding pair. In Arizona, where the species has been studied in detail (e.g., Brown, 1970, 1974; Brown and Brown, 1981), extended families of jays defend a single territory from neighboring clans. Within clans, certain individuals may spend their entire lives without emigrating, eventually breeding in the same territory where they originated and helped, even while their parents remain alive.

Rarely, Florida Scrub Jays also exhibit plural nesting. The phenomenon occurs when boundaries between related pairs are slow to develop following territorial budding. In addition, certain populations of Mexican Jays exhibit singular nesting typical of the Florida population (J. L. Brown, pers. comm.). Thus, the gap between singular and plural nesting is bridged in several ways among extant populations within the genus (Figure 6-1). The significance of these transitional stages is stressed in the final section of this chapter.

As elaborated below, suboptimal habitat is almost non-existent for Florida Scrub Jays. Within usable habitat, the density of breeding pairs is extremely stable (Figure 6-2), a pattern also noted in the canyon-bottom populations of Mexican Jays studied by Brown and Brown (1984). During 14 years of observation (1971–1984), breeder density in our population has varied from 5.8 to 8.1 pairs per 40 ha, showing a coefficient of variation of only 8%. This level of stability in a breeding population is unequaled in any other land bird yet measured (Ricklefs, 1973). Besides the breeders, an excess of *potential* breeders always exists in our Florida Scrub Jay population (Figure 6-2). The excess includes juveniles up to age one year and helpers one to seven years old. The latter category in particular represents jays that have reached physiological maturity and are capable of breeding. The Florida Scrub

FIG. 6-2 Population densities of breeders, older helpers (>1 year old), and juveniles within the marked study population of Florida Scrub Jays, 1971–1984. Density measurements are plotted for January, April (breeding), July, and October each year. Uppermost line represents the total population density, with annual peaks reflecting the July infusion of new independent juveniles from the breeding season.

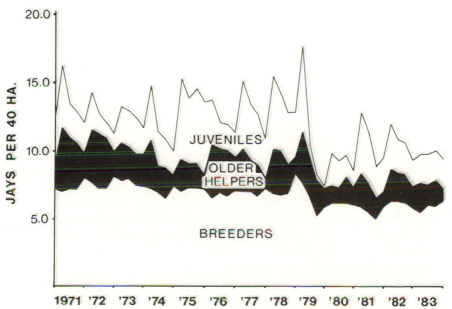

Jay social system therefore consists of constant interplay between different social classes of jays, as non-breeders work their way into the saturated breeding population and new potential breeders are produced each year.

DEMOGRAPHY OF A MARGINAL HABITAT

In Florida, the Scrub Jay is restricted to the oak scrub, a rare and patchy habitat dominated by several species of stunted oaks and other woody perennials (detailed descriptions in Abrahamson et al., 1984; Woolfenden and Fitzpatrick, 1984). This distinctive plant formation appears only on modern or ancient sand dunes found along recent or Pleistocene seashores. Abrupt ecotones exist between the scrub community and the other, more widespread Florida habitats. These ecotones reflect the borders of well-drained sandy soils, upon which the oak scrub is dependent. Therefore, habitat that is acceptable to Florida Scrub Jays is sharply bounded by vast expanses of habitat that is totally unusable.

Some variation in structure and quality does exist within the oak scrub habitat, although it is not as variable as oak-dominated habitats in western North America. In Florida, the most important natural variation occurs as a result of site-to-site differences in fire histories. The scrub is a fire-maintained habitat. Its low, open profile (1 to 2 m high) persists through the periodic occurrence of lightning fires. Left unburned, the scrub acquires a higher profile (reaching 5 m or more), accumulates a dense leaf litter, and contains ever-increasing densities of tall pine trees. This succession proceeds gradually toward habitat that is favored by an arboreal competitor, the Blue Jay *(Cyanocitta cristata)*, a species that shuns the more open patches of scrub.

Through human intervention, several patches of scrub adjacent to our main study tract have gone unburned since the mid-1930s. These areas once harbored numerous Scrub Jay territories, but now are used only sporadically. During 15 years we have monitored the lives and breeding performances of those jays that did attempt to occupy these "marginal" areas of tall, dense scrub. Table 6-1 compares survivorship and reproduction in these sporadically used areas with similar measures in the main study tract, most of which has burned several times since the 1930s.

Both in reproductive output and in survivorship, jays using the taller habitat show significantly poorer success. Despite more numerous nest attempts they produce fewer fledglings in the average year, because of higher rates of nest predation. Those fledglings that are produced also suffer significantly higher mortality in the tall habitat. Finally, the breeders themselves suffer higher mortality in the tall habitat than do those in the open scrub. These three factors combine to produce vastly

TABLE 6-1   Mean survivorship and reproductive success of Florida Scrub Jays in two kinds of scrub habitat, 1969–1984.

	Scrub quality	
	Low, open, periodically burned	Tall, dense, unburned
N *(pairs)*	326	52
Annual success		
Seasonal nest attempts	1.3	1.5
Fledglings per pair	2.07*	1.60
Juvenile survival (first year)	0.321*	0.193
Breeder survival (annual)	0.806**	0.698
Expected lifetime success per individual		
Breeding seasons	4.9	3.3
Fledglings	5.1	2.7
Yearlings	1.6	0.5
Breeding offspring	1.1	0.4

$*p < .05$
$**p < .01$

lower expected lifetime fitness for jays that settle to breed in the tall, unburned scrub habitat (.51 vs. 1.63 yearlings produced per individual in an average lifetime).

This remarkable difference in expected fitness between two adjacent oak scrub formations, which differ principally in the dates of their most recent fires, underscores the extreme degree of habitat specificity in the Florida Scrub Jay. Indeed, we repeatedly have witnessed behavioral evidence that the jays attempt to avoid the dense zones altogether, or engage in active combat with neighbors in order to include a piece of open scrub in their territories (Woolfenden and Fitzpatrick, 1984). Only sporadically do individuals establish territories in the areas of tall scrub.

The above results prompt us to propose here a simple demographic model for the conditions under which suboptimal habitat should be avoided. Such a question is relevant to cooperative breeding because it translates directly into the terms of Koenig and Pitelka's (1981) "resource localization" model, as discussed below. Consider two habitats, a good one and a poor or marginal one. Expected values for lifetime fitness of individuals occupying these habitats *once they are established as breeders* may be represented by R_g and R_p, respectively. (These two values are shown for Florida Scrub Jays in Table 6-1, lower two rows.) For each habitat we must consider a second variable, Ψ, representing the probability that a yearling (or any other pre-reproductive we choose) can eventually become established as a breeder in that habitat; Ψ always will be less than 1.0, but in little-used habitats it can approach 1.0. For a potentially dispersing yearling choosing between these two habitats, and assuming that the choice is final, if

$$R_g \Psi_g > R_p \Psi_p \tag{1}$$

then lifetime fitness will be greater in the good habitat. The poor habitat should be shunned, even if it offers a comparatively high probability of breeding (i.e., Ψ_p near 1.0). Equally, if

$$\Psi_g > R_p/R_g \tag{1'}$$

then even if $\Psi_p = 1.0$ the poor habitat is not worth settling into. Instead, the individual should do whatever it must, including delaying breeding if necessary, in order to settle in and breed in the good habitat. It is important to note that Ψ is an overall lifetime probability, not necessarily an annual one. In cases of delayed breeding it sums all annual probabilities of becoming a breeder, and includes each year's risk of death before reproduction according to the following formula (Woolfenden and Fitzpatrick, 1984):

$$\Psi = B \cdot \sum_{i=1} l_h^i \cdot (1 - B)^{i-1}$$

where B is the *annual* probability of becoming a breeder, l_h is annual survivorship before breeding, and i is measured in years.

Substituting the actual values for Florida Scrub Jay fitness in two habitats (Table 6-1) into expression (1') yields the conclusion that in our study area yearlings should not disperse into the poorer habitat if their overall chance of locating or creating a breeding position in the better habitat (Ψ_g) is greater than $(.51/1.6) = 0.32$. Indeed, we have independently calculated that the overall Ψ in our population is about 0.62 for yearlings (see below).

SURPLUS OF POTENTIAL BREEDERS

Habitat limitation is not in itself sufficient to explain the persistence of a backlog of non-breeding individuals. In demographic terms, such a backlog represents its own age-structured population of recruits, some of which are "lost" each year through death or by becoming breeders. Annual recruitment into the backlog is achieved through reproduction by the breeders, followed by survival of juveniles to potential breeding age.

For Florida Scrub Jays, potential breeding age is reached on the first birthday, at which point we refer to them as yearlings. Many Scrub Jays in western North America breed at this age (Ritter, 1983), as have a

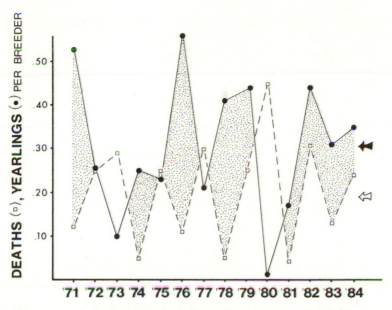

FIG. 6-3 Annual production of yearlings (dots) compared with death rates of breeders (open squares) over 14 consecutive years in the marked population of Florida Scrub Jays. Mean values for both measures are indicated by arrows (black = yearlings per breeder, open = deaths per breeder). Stippled areas illustrate those years in which production of potential breeders exceeded the rate at which breeding vacancies appeared. In 1980 an apparent epidemic caused heavy mortality among both breeders and juveniles.

very few Florida Scrub Jays (Woolfenden and Fitzpatrick, 1984). Therefore, we consider yearlings to represent the class of individuals that have entered the competition for breeding space within our population.

Figure 6-3 graphs the annual recruitment of potential breeders (i.e., yearlings, not fledglings) within our marked population between 1971 and 1984. With few exceptions (N = about 5 immigrants in 15 years) all yearling recruits represent known-parentage jays marked by us as nestlings. Successful emigration from our tract by marked juveniles also is essentially nil, so that Figure 6-3 can be assumed to depict the true annual variation in per capita recruitment within the population at large. Graphed on the same scale in Figure 6-3 is the annual death rate of breeders within the same population over the same period. In a stable population breeder deaths represent vacancies that can be filled from the standing pool of potential breeders. If breeder death rates and yearling production were about equal, little competition would exist for breeding space, even though the young have no peripheral habitats to disperse into. However, these values are not equal (Figure 6-3, arrows). In 9 of 14 years a net surplus of yearlings was produced, and yearling production roughly equalled breeder mortality in 2 additional years. Only in 3 years (1972–73, 1976–77, 1979–80) did breeder deaths exceed the production of potential replacements. (In one of these years, 1979–80, an epidemic swept through the population, killing all but one marked juvenile and 45% of the breeders; see Woolfenden and Fitzpatrick, 1984, for details and results of this "natural experiment.")

A net excess of potential breeders, and no marginal habitat to support them, prompted us to propose a model for the evolution of cooperative breeding through breeding-space competition (Woolfenden and Fitzpatrick, 1984). According to our model, individuals realize greater lifetime fitness by delaying dispersal and helping under the following demographic conditions:

$$D_0 < \frac{l_1 \cdot k}{R} + l_1 \cdot B \cdot \sum_{i=1} l_h^i \cdot (1 - B)^{i-1} \qquad (2)$$

where: D_0 = probability of successful dispersal before age 1 year

l_1 = survival to age 1 year in the natal territory

k = increment of additional reproduction of siblings produced through helping, corrected for average relatedness to those offspring

R = expected lifetime reproduction once breeding status is attained

B = average annual probability of obtaining breeding space by remaining in the natal territory

l_h = annual survival as a helper beyond age 1 year

For present purposes, we emphasize only the general notion behind this model (see Woolfenden and Fitzpatrick, 1984, for detailed discussion). The model assumes that some probability, B, always exists that an individual can successfully become a breeder without actively dispersing. Individuals can, of course, affect their own B, and B can increase with age (see below). The criterion for "choosing" to disperse rather than to stay at home, then, depends upon the relationship between this probability, plus any kinship benefits gained through helping, and the probability of successful, permanent dispersal early in life (D_0). We contend that D_0 is a demographic parameter of utmost importance in the evolution of cooperative breeding. Its significance has been alluded to, but not stressed, in several recent reviews of the topic (e.g., Brown, 1978; Emlen, 1982a, 1982b; Koenig and Pitelka, 1981).

In the Florida Scrub Jay, we cannot directly measure D_0, because essentially no jays attempt dispersal before their first birthday. However, demographic data on the existing population allow us to estimate what it might be under present conditions if they did try to disperse.

In the following derivation D_0 is estimated simply by dividing the average annual number of breeding vacancies by the average number of potential recruits that could fill those vacancies. Recruits consist of yearlings plus any backlog population of older, non-breeding individuals left over from preceding years. Where m = birth rate (annual fledgling production per individual breeder), l_1 = juvenile survival to age 1 year, l_h = survival of older non-breeders (helpers), and l_B = survival of breeders:

$$D_0 = \frac{(1 - l_B)}{m \cdot l_1 + \gamma \cdot l_h} \qquad (3)$$

The new variable, γ, represents the "standing crop" of older non-breeders. In essence, it is an *index of breeding-space competition*, because it directly measures the relative number of competitors that persist in the population between successive breeding seasons. Where γ is zero, yearlings face only other yearlings in the competition for space, and the losers die. If, instead, the losers survive at some annual average, l_h, then γ increases until an equilibrium is reached where

$$\gamma_{eq.} = (l_h \cdot \gamma_{eq.} + m \cdot l_1) - (1 - l_B) \qquad (4)$$

That is, the theoretical "standing crop" represents the equilibrium number of non-breeders (expressed per breeder) at which losses through death $(1 - l_h)$ are exactly replaced by new, left-over yearlings $(m \cdot l_1)$ after each year's breeding spaces $(1 - l_B)$ are filled. Note that average group size, among monogamous and singular nesting species, is simply equal to $\gamma + 2$. In this light, group size emerges as a demographic result, not as the strategic optimum that it often is considered to be (e.g., Bertram, 1978; Koenig, 1981).

Expression (4) can be solved for γ:

$$\gamma = \frac{m \cdot l_1 - (1 - l_B)}{1 - l_h} \qquad (5)$$

Substituting expression (5) into expression (3) yields the following derivation of D_0:

$$D_0 = \frac{1 - l_B}{m \cdot l_1 + l_h \left[\dfrac{m \cdot l_1 - (1 - l_B)}{1 - l_h} \right]}$$

$$= \frac{(1 - l_B)(1 - l_h)}{(1 - l_h) \cdot m \cdot l_1 + l_h \cdot m \cdot l_1 - l_h \cdot (1 - l_B)}$$

$$= \frac{(1 - l_B)(1 - l_h)}{m \cdot l_1 - (1 - l_B) \cdot l_h} \qquad (6)$$

Thus, the theoretical probability that a yearling would succeed at becoming a breeder by dispersing early (before age one year) can be calculated directly from four simple demographic measurements: birth rate (m) and survival rates for three social classes (l_1, l_h, and l_B).

The above calculation uses only population mean values for demographic parameters, taking no account of variances in survival and reproduction. Variance can affect the net rate of appearance of breeding vacancies. More importantly, the simple calculations above assume that yearlings and older non-breeders enter the breeding space competition as if it were a simple lottery—each individual having an equal chance of winning. As we discuss below, this probably is not true in real-world cases. (Indeed, the opportunity to improve one's chances of winning represents an important component to the strategy of remaining in the natal territory.) Nevertheless, expression (6) does provide a first-order means for comparing the degree of breeding-space competition among different populations. Inserting the actual demographic values for the Florida Scrub Jay into expression (6) yields:

$$D_0 = \frac{(0.194)(0.28)}{(1.04)(0.33) - (0.194)(0.72)} = 0.27$$

Therefore, as a demographic maximum estimate, yearling Florida Scrub Jays would realize about a one-in-four success rate if they were to attempt dispersal before the end of their first year.

Numerous factors reduce the realized D_0 below the theoretical maximum just calculated. Most important among these are (1) age-specific costs to attempting dispersal, such that early entry into the "lottery" is penalized more than later entry; (2) any combination of processes by which older non-breeders realize improved success at winning breeding slots (e.g., dominance effects, group size benefits, improved ability in overt competition, etc.); (3) any demographic perturbation that increases γ, thereby raising the total number of entries into the lottery. As discussed below, the first two in this set of factors clearly apply to the Florida Scrub Jay breeding population.

AGE-SPECIFIC SURVIVAL OF NON-BREEDERS

Even when living in the relative safety of the home territory, mortality among young Florida Scrub Jays is high during the first 12 months of life. Only one-third of all fledglings become yearlings, and in some years the fraction is near zero. In Figure 6-4 we graph juvenile survivorship from fledging to age 12 months, showing the overall mean values (1969–1984), and also the separate curves for 1979 (the year of the epidemic) and all years excluding the disastrous one.

If young Florida Scrub Jays were to disperse a few months after fledging, as in the western populations, we assume that their survivor-

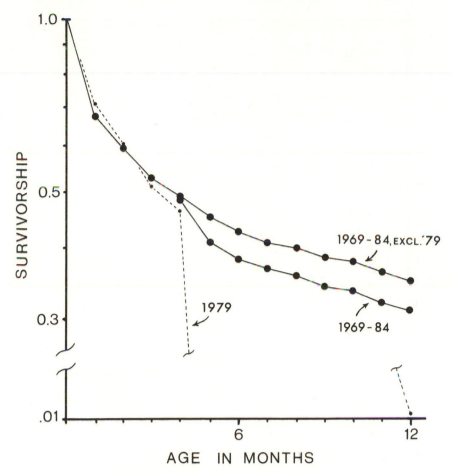

FIG. 6-4 Average survivorship during the first 12 months of life among 772 Florida Scrub Jay fledglings. Because the epidemic caused anomalous survivorship during one year (1979–1980, dotted line), we illustrate the average curves both including and excluding that year.

ship could only be worse than the 31% shown in Figure 6-4. We suspect that it would be substantially worse, primarily because of their narrow habitat tolerance already discussed. To generalize from this result, delayed dispersal becomes an increasingly favorable strategy as juvenile mortality and the juvenile "learning period" become greater.

We can directly test our assumption that active dispersal increases the mortality of young jays. We make use of the different dispersal strategies of male versus female helpers, with males acting as our "control" group. As shown in Figure 6-5, it takes young jays in Florida nearly a full year to lower their mortality approximately to the level of breeding adults (19%). For males, who remain in the home territory until becoming breeders, this low mortality stabilizes near that of breeders (Figure 6-5). For females the story is different. During the

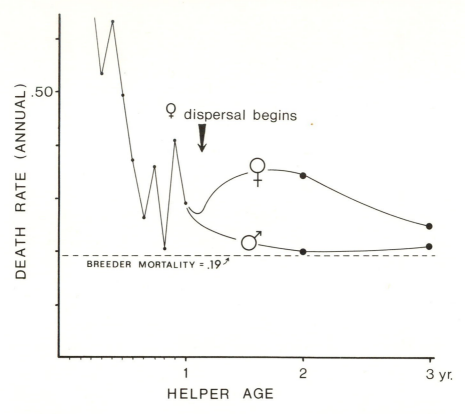

FIG. 6-5 Mortality estimates, converted to their annual rates, over the first three years of life among non-breeding Florida Scrub Jays. Active dispersal forays by female yearlings cause their mortality to rise, while male helpers approach the low rate of mortality of breeding jays (dotted line).

next year or two, females make frequent dispersal forays away from the home territory in search of breeding vacancies, sometimes staying away for days at a time. In so doing, they locate breeding slots more quickly, but they also experience a sharp increase in their death rate. Compared to the "control" males that only rarely leave home as yearlings, females' dispersal tactics cost them by placing them at a significantly higher risk of death (Figure 6-5). Presumably, as jays gain experience, their risks in occasionally departing from home get progressively lower. Therefore, we surmise that this risk would be even greater among any individuals that attempt to disperse earlier than age one year.

AGE-SPECIFIC INCREASE IN BREEDING PROBABILITY

Earlier, we postulated that helpers could gain directly by helping to raise additional family members, in large part because the helpers

TABLE 6-2   Fates of older male helpers and presence or absence of younger helpers. Samples represent helper-years; individual helpers are tallied separately for each year they began as an older helper, age two or older.

	Buds or inherits territory	Does not bud or inherit territory		
		Breeder replacement	Still helping	"Dead"
With younger helper (N = 92)	31 (.34) $p < .0005$	16 (.17)	38	7 (.08)
Without younger helper (N = 69)	9 (.13)	10 (.14)	39	11 (.16)

might stand a better chance of gaining a breeding space, often inherited through territorial budding (Woolfenden and Fitzpatrick, 1978). Subordinate helpers could help increase the family's territory size, raising the likelihood of the dominant (or the subordinant, later) inheriting part of the territory as his own. Data are now sufficient to test our proposition. Table 6-2 shows that, as predicted, a significantly higher proportion of older males inherit or bud a territory when subordinates are present (34%) vs. absent (13%). Overall, older helpers stand almost twice the probability of breeding the following year if they have subordinate helpers in the group (51 vs. 27%, Table 6-2). A direct mechanism therefore exists by which male helpers can improve their own breeding opportunities without leaving home.

A few females also inherit breeding space. However, most females actively attempt to disperse from the natal territory before age three. This behavior results in an increasing probability of breeding among the older female helpers (Figure 6-6).

Table 6-2 also indicates a trend toward increased survivorship among male helpers with subordinates. The difference in annual death rates (8 vs. 16%) is not significant, however.

The overall probabilities of becoming breeders the following year are summarized in Figure 6-6 for each Florida Scrub Jay age and sex class. Virtually no fledgling breeds at age one year. Yearling males and females experience probabilities between 0.3 and 0.4, depending on whether they have siblings in the family during their second year. Older helpers show a graded series of probabilities, with the high rate for females (.71) reflecting their active dispersal strategy (virtually all the remainder die). Males with siblings show higher probabilities than those without, and dominant older males show the highest annual probability for their sex.

These data indicate that delayed dispersal can be accompanied by steadily improving chances of becoming a breeder. Where this holds true, early dispersers actually are penalized—their probability of success is even lower than the theoretical D_0 calculated above, which assumed that all potential dispersers had an equal chance of winning a

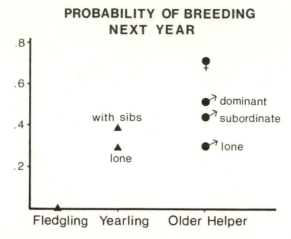

FIG. 6-6 Average probabilities of breeding the following year, for seven categories of non-breeding Florida Scrub Jays. Sexes are combined for fledglings and yearlings (triangles), and separated for older helpers (age two or older).

breeding slot. Therefore, a system of positive feedback further enforces delayed breeding, especially when individuals can increase their own breeding opportunities by helping.

DELAYED DISPERSAL AS EXTENDED PARENTAL INVESTMENT

The currency of fitness in a wild population is offspring (e.g., Grafen, 1982), and the only biologically meaningful offspring are those that become breeders themselves. Our first analysis of this crucial measure currently is in preparation, but a dramatic preliminary result already is apparent and can be reported here. Of exactly 100 Florida Scrub Jay breeders whose entire breeding histories we followed (lasting 1 to 11 breeding seasons, mean = 3.53), 46% never produced even a single yearling, much less a breeding descendant. Fully 54% of the breeders do not produce breeding offspring, and fewer than 25% of the breeders produce 75% of the breeding recruits within a Florida Scrub Jay population. We have observed jays breeding as many as 8 years in succession without raising a single yearling, although a strong correlation does exist between breeding seasons attempted and number of yearlings produced. The main causes of such high variance in lifetime fitness are high rates of nest predation among new breeders, the higher mortality of breeders without helpers, and the high variance in success rates of yearlings at becoming breeders themselves (Fitzpatrick and Woolfenden, in press).

In a population where most adults utterly fail to reproduce, success at producing even a few breeding offspring insures higher fitness than most other individuals will achieve. Investment by parents in their off-spring should be directed preferentially to those in which the invest-ment can make the biggest increase in ultimate probability of these off-spring breeding. Such a probability can be expressed in terms of reproductive value. Figure 6-7 graphs the reproductive value of Florida Scrub Jay offspring at various stages in their development. (An average new breeder produces about 16 eggs in an average lifetime, and about one egg in 16 becomes another breeder.) For present purposes, the key feature of the curve in Figure 6-7 is the relatively high reproductive values of yearling and two-year-old non-breeders. From the point of view of the parents' ultimate reproductive success, these older individ-uals are "worth" seven to ten times the value of an egg or nestling.

In most birds, older offspring are expelled from the home territory. However, what would be the result of expulsion for birds such as Flor-

FIG. 6-7 Reproductive value, or expectation of future reproduction, for Florida Scrub Jays, calculated at different developmental periods. Each egg, on the average, exactly replaces itself; hence it has a reproductive value of 1.0. An average new breeder eventually produces about 16 eggs in a lifetime; some offspring breed at age two (dashed line ascending from yearling's value). An average nest contains about 3.4 eggs; hence the value of a clutch or nestling brood (open squares) is proportionately higher than that of an individual (closed squares).

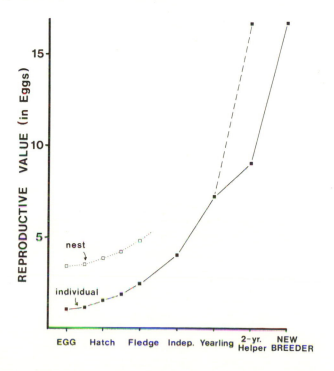

ida Scrub Jays? Recall that all usable habitat is already saturated, and that marginal habitat is too poor to support dispersing jays. Expulsion of yearling Florida Scrub Jays would result in vastly increasing the risk that these valuable offspring, who have succeeded in surviving nearly to breeding age, would perish before locating a breeding slot. As shown in Figure 6-7, a single yearling represents a higher reproductive value than an entire nestful of the present year's offspring. Therefore, retention of yearling and older offspring often will be favored *even at the expense of lowered success* in immediate nest attempts.

If remaining in the natal territory is in the best interests of the off-spring as a means of becoming a breeder, then normally the best inter-ests of the parents are served by permitting this behavior. This view of delayed breeding is in direct contrast to the competition-oriented view expressed by Emlen (1982a, 1982b) and Emlen and Vehrencamp (1983). Our interpretation likens retention of young to a form of "extended parental care" (Ligon, 1981), in which the ultimate interest of the parents lies in maximizing the number of breeding *grand-off-spring*, not simply increasing the number of direct offspring. Such a strategy in many circumstances can be expressed by investing more in older surviving offspring than in new younger ones. As we emphasized elsewhere (Woolfenden and Fitzpatrick, 1984), in an age-structured population of long-lived breeders, the interests of offspring and par-ents in this regard converge to become nearly identical. The result is a breeding system that conforms to the simplest definition of coopera-tion, namely, that all parties gain from the shared activities.

The idea that retention of helpers represents a form of parental care is not a new one. We implied the notion in Woolfenden and Fitzpatrick (1978), following Zahavi (1974). Ligon (1981) stated it directly in his summary of the social organization of the Green Woodhoopoe *Phoen-iculus purpureus*. More recently, Brown and Brown (1984) arrived at a similar interpretation (which they term "parental facilitation") of delayed dispersal and offspring philopatry in the Mexican Jay (see below). The conceptual advance that this interpretation represents over earlier ones (e.g., Brown, 1974, 1978; Emlen, 1982b) lies in divorcing the existence of helpers from any payment or direct benefit they must contribute to the breeders in order to be allowed to remain. That they represent potential breeders may be all that is required for parents to continue tolerating them, if the alternatives are sufficiently bleak.

EVOLUTION OF COOPERATION IN APHELOCOMA

We return to the general question of the evolution of differing degrees of group living and cooperative breeding in the genus to which the Florida Scrub Jay belongs (Figure 6-1). For convenience, we refer to complex social systems as if they are derived from more simple ones. We emphasize, however, that the reverse could just as easily be true in individual cases. New World jays are predominantly highly social (e.g., Brown, 1974). Therefore, the simple sociality of Scrub Jays in western North America might represent a reversion to a more primitive avian system, a demographic change caused by release from the ecological restrictions that characterize its more cooperative relatives. For this reason the social "stages" in Figure 6-1 are connected by lines and not by arrows. We consider the transitions between these stages to be reversible.

Delayed breeding, increasing group size, increasing philopatry, and finally the retention of offspring at home as breeders represent a series of life history strategies associated with decreasing access to breeding opportunities for potentially dispersing young. Presumably the common denominator within *Aphelocoma* has been some limitation in habitat—especially the relative absence of spatial and temporal variation in habitat quality. (We note, however, that habitat limitation has yet to be demonstrated for any population outside of Florida.)

For monogamous birds, a broad range of demographic conditions exists in which individuals maximize fitness simply by attempting to breed at the earliest possible time in their lives. Relatively high adult mortality as well as ample spatial variation in habitat quality probably are the principal features that make early post-natal dispersal the best strategy for most birds. These conditions result in reliably high probabilities of successful breeding at age one. Furthermore, these are precisely the population characteristics that place a competitive premium on early breeding (Lewontin, 1965). The Scrub Jays of western North America exhibit this common, "primitive" condition.

When the production and survival of potential dispersers exceeds the appearance of breeding vacancies, delayed breeding is forced upon at least a few individuals. Vacancies can become scarce either through reduced availability or quality of marginal habitats, or a lowered rate of appearance of new, usable habitat patches. Even if the absolute appearance rate of vacancies remains stable, competition for them can increase through increased production or survival of the potential recruits. If these recruits can survive in marginal habitats, and in so doing achieve a sufficient probability of locating a breeding space early, then dispersal from the natal territory is favored even when many individuals must delay breeding. The Scrub Jays of Santa Cruz Island, California, exhibit this peculiar, intermediate condition (Atwood, 1980).

After delayed breeding, the next major stage of evolution toward cooperative breeding is the appearance of delayed dispersal from the natal territory. Two separate behavioral traits must evolve: (1) reduced propensity for the offspring to leave their parents' territory, and (2) parental tolerance of older offspring remaining in their territory. These traits are both favored when the probabilities and rewards of successful early dispersal fall below those gained by staying home. Such conditions can accompany reduced survival of pre-reproductive dispersers, increased survival of breeders, increased competitive advantages gained by delaying dispersal, kinship benefits gained through helping while at home, or combinations of all these factors. Because parental fitness is maximized only by maximizing the production of *breeding* offspring, tolerance of their offspring within the natal territory becomes a reproductive strategy in its own right, a form of extended parental care. The Scrub Jays of Florida exhibit this stage of social evolution.

Helping behavior arises in conjunction with delayed dispersal from the natal territory. Typically, offspring care by helpers closely resembles that of breeders, and it remains unclear to us that any selective advantage need exist for parental behavior to be practiced by pre-reproductive offspring. We have suggested the possibility that this behavior occurs as an innate response to the presence of begging young within the territory (Woolfenden and Fitzpatrick, 1984, in press). In any case, the behavior certainly is reinforced where it results in any increase in the inclusive fitness of the helper, either through increased production of collateral kin or through increases in the probability of acquiring breeding status. Both advantages appear to exist among the New World jays that have been studied in detail, and we have presented a case for the latter advantages being especially important in the genus *Aphelocoma* (Woolfenden and Fitzpatrick, 1984).

Once delayed dispersal becomes the rule in a population, average group size is a simple demographic product of birth rates and survivorship of juveniles, helpers, and breeders. We already have shown that group size varies from year to year according to annual variations in these four measures (Woolfenden and Fitzpatrick, 1984). Long-term changes in the average value of any one measure will affect the long-term, average group size in the same way. Of course, the same group size can result from various combinations of demographic values. This concept is illustrated graphically in Figure 6-8, where we employ the formula for calculating average group sizes, derived in a preceding section.

According to J. L. Brown (pers. comm.) certain populations of Mexican Jays in northeastern Mexico exhibit single-nest breeding among groups that average about five to six individuals per group, about twice the size of Florida Scrub Jay groups. We predict that the social system of these populations otherwise closely resembles that of the Florida Scrub Jay, and that some minor, measurable demographic difference accounts for the larger group sizes.

An intriguing point is illustrated in Figure 6-8 regarding the size and structure of cooperative-breeding family units. If we assume realistic upper limits to reproduction and annual survival, then the *maximum possible* average group size in a monogamous population will be between five and eight individuals. Larger groups can occur upon occa-

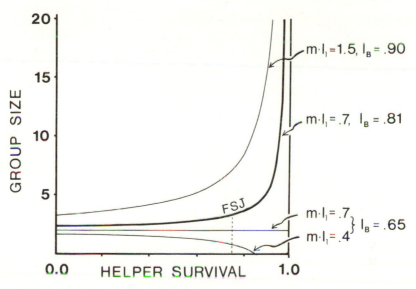

FIG. 6-8 Results of four simulations calculating equilibrium group sizes according to the formulas: group size $= 2 + \gamma$, and $\gamma = [m \cdot l_1 - (1 - l_B)]/1 = l_h$, as derived in the text. Group size is plotted against annual helper survival, l_h, which can vary from 0 to 1.0. The heavy line represents the approximate values for yearling production ($ml_1 = 0.7$) and breeder survival ($l_B = 0.81$) in our Florida Scrub Jay population. An average helper survivorship of about 0.76 (combining males and females) yields an average group size of about 3.0, which matches the observed value in our population. Because realistic populations rarely would show average survival as high as 0.9 or yearling production as high as 1.5, group sizes averaging higher than five to eight individuals are demographically difficult to achieve for singular-nesting species, and are rare in nature.

sion, but to show an *average* group size of more than about eight individuals the population would have to enjoy unusually high reproduction and juvenile survival (averaging more than 1.5 yearlings per pair) combined with spectacularly high average survival among the helpers and breeders (more than .90). Few avian species meet these requirements. No singular-nesting, cooperative-breeding species known to us averages more than eight individuals per group. However, the genus *Aphelocoma* contains at least one population, the Mexican Jays of Arizona, in which group size averages between ten and fifteen individuals. However, instead of representing a counter-example, this population contains a major new step in the same social continuum we have been describing. The new stage is the joining together of separate monogamous units into a single, plural-nesting group. In having two or three nesting pairs within the territory, Mexican Jay units circumvent the demographic constraints on group size illustrated in Figure 6-8.

We concur with Brown (1974) in supposing that the evolution of larger groups in the Mexican Jay stems from competitive advantages enjoyed by larger groups over smaller ones when usable space is scarce. Moreover, the territorial budding system of the Florida Scrub Jay (Woolfenden and Fitzpatrick, 1978, 1984) represents the logical intermediate step from which plural nesting has evolved. In Florida, competition for space is not so severe as to prohibit simple pairs or trios from carving out their own territories through budding. For this reason, the occasional "double families" we have witnessed in Florida show no higher success at reproduction or territory maintenance than do smaller groups.

We propose that in cases where demographic pressure reaches the point where the occasional fusion of units does result in measurably higher or more lasting success, then the budding process will begin to give way. This could happen if larger units, formed through incomplete budding, more frequently succeed at maintaining territories than do smaller units. Formation of boundaries between parents and their budding offspring therefore becomes less advantageous. The result would be the gradual spread of lineages in which some offspring never depart, but rather become breeders within the natal unit. Viewed in this way, plural nesting represents an evolutionary adaptation. Its chief advantage lies in increasing the size of the cooperating unit beyond that which is demographically possible if only one pair is allowed to breed in the group. The form of plural nesting exhibited by many tropical jays therefore can be interpreted as an advanced form of parental care, in which both offspring and parents derive competitive advantages through insuring that those offspring will have a place to breed and produce offspring of their own.

ACKNOWLEDGMENTS

We are especially grateful to the Archbold Biological Station and its staff for their continued cooperation in making our field studies possible. We thank J. D. Ligon and F. Pitelka for their comments on the manuscript.

LITERATURE CITED

Abrahamson, W. G., A. F. Johnson, J. N. Layne, and P. A. Peroni. 1984. The vegetation of the Archbold Biological Station, Florida: An example of the southern Lake Wales Ridge. *Florida Scientist* 47:209–250.

Atwood, J. L. 1980. Social interactions in the Santa Cruz Island Scrub Jay. *Condor* 82:440–480.

Bertram, B. C. 1978. Living in groups: Predators and prey. In: Krebs, J. R., and N. B. Davies (eds), Behavioural Ecology: An Evolutionary Approach. Sinauer, Sunderland, Mass., pp. 64–96.

Brown, J. L. 1970. Cooperative breeding and altruistic behavior in the Mexican Jay, *Aphelocoma ultramarina. Animal Behaviour* 18:366–378.

Brown, J. L. 1974. Alternate routes to sociality in jays—with a theory for the evolution of altruism and communal breeding. *American Zoologist* 14:63–80.

Brown, J. L. 1978. Avian communal breeding systems. *Annual Review of Ecology and Systematics* 9:123–156.

Brown, J. L., and E. R. Brown. 1981. Extended family system in a communal bird. *Science* 211:959–960.

Brown, J. L., and E. R. Brown. 1984. Parental facilitation: Parent-offspring relations in communally breeding birds. *Behavioral Ecology and Sociobiology* 14:203–209.

Emlen, S. T. 1978. The evolution of cooperative breeding in birds. In: Krebs, J. R., and N. B. Davies (eds), Behavioural Ecology: An Evolutionary Approach. Sinauer, Sunderland, Mass., pp. 245–281.

Emlen, S. T. 1982a. The evolution of helping behavior. I. An ecological constraints model. *American Naturalist* 119:29–39.

Emlen, S. T. 1982b. The evolution of helping behavior. II. The role of behavioral conflict. *American Naturalist* 119:40–53.

Emlen, S. T. 1984. Cooperative breeding in birds and mammals. In: Krebs, J. R., and N. B. Davies (eds), Behavioural Ecology: An Evolutionary Approach. Sinauer, Sunderland, Mass., pp. 305–339.

Emlen, S. T., and S. L. Vehrencamp. 1983. Cooperative breeding strategies among birds. In: Brush, A. H., and G. A. Clark, Jr. (eds), Perspectives in Ornithology. Cambridge Univerisity Press, Cambridge, pp. 93–120.

Fitzpatrick, J. W., and G. E. Woolfenden. In press. Components of lifetime fitness in the Florida Scrub Jay. In: Clutton-Brock, T. H. (ed), Reproductive Success. University of Chicago Press, Chicago.

Gaston, A. J. 1978. The evolution of group territorial behavior and cooperative breeding. *American Naturalist* 112:1091–1100.

Grafen, A. 1982. How not to measure inclusive fitness. *Nature* 298:425–426.

Koenig, W. D. 1981. Reproductive success, group size, and the evolution of cooperative breeding in the Acorn Woodpecker. *American Naturalist* 117:421–443.

Koenig, W. D., and F. A. Pitelka. 1981. Ecological factors and kin selection in the evolution of cooperative breeding in birds. In: Alexander, R. D., and D. Tinkle (eds), Natural Selection and Social Behavior: Recent Research and New Theory. Chiron, New York, pp. 261–280.

Lewontin, R. C. 1965. Selection for colonizing ability. In: Baker, H. G., and G. L. Stebbins (eds), The Genetics of Colonizing Species. Academic Press, New York, pp. 77–94.

Ligon, J. D. 1981. Demographic patterns and communal breeding in the Green Woodhoopoe, *Phoeniculus purpureus.* In: Alexander, R. D., and D. Tinkle (eds), Natural Selection and Social Behavior: Recent Research and New Theory. Chiron, New York, pp. 231–243.

Reyer, H-U. 1980. Flexible helper structure as an ecological adaptation in the Pied Kingfisher *(Ceryle rudis). Behavioral Ecology and Sociobiology* 6:219–229.

Ricklefs, R. E. 1973. Fecundity, mortality, and avian demography. In: Farner, D. S. (ed), Breeding Biology of Birds. National Academy of Science, Washington, pp. 366–435.

Ritter, L. V. 1983. Nesting ecology of Scrub Jays in Chico, California. *Western Birds* 14:147–158.

Vehrencamp, S. L. 1979. The roles of individual, kin, and group selection in the evolution of sociality. In: Marler, P., and J. G. Vandenbergh (eds), Social Behavior and Communications. Handbook of Behavior and Neurobiology, vol. 3. Plenum Press, New York, pp. 351–394.

Verbeek, N. A. M. 1973. The exploitation system of the Yellow-billed Magpie. *University of California Publications in Zoology* 99:1–58.

Woolfenden, G. E., and J. W. Fitzpatrick. 1978. The inheritance of territory in group-breeding birds. *BioScience* 28:104–108.

Woolfenden, G. E., and J. W. Fitzpatrick. 1984. The Florida Scrub Jay: Demography of a Cooperative-Breeding Bird. Princeton University Press, Princeton, N.J.

Woolfenden, G. E., and J. W. Fitzpatrick. In press. Sexual asymmetries in the life history of the Florda Scrub Jay. In: Rubenstein, D., and R. W. Wrangham (eds), Ecology and Social Evolution: Field Studies of Birds and Mammals. Princeton University Press, Princeton, N.J.

Zahavi, A. 1974. Communal nesting by the Arabian Babbler: A case of individual selection. *Ibis* 116:84–87.

7

Parent-Offspring Interactions in Anthropoid Primates: An Evolutionary Perspective

Jeanne Altmann

FROM an evolutionary viewpoint what is special about parent-offspring interactions? Most behavior that is performed at a reproductive or survival "cost" by one animal and that "benefits" (both terms sensu Hamilton, 1964) a recipient by increasing its chances of survival is thought to be done at some direct, usually immediate, cost to the actor. Generally such behavior will be selected against unless the cost is recouped, either through reciprocity or through inclusive fitness effects. The case of behavior directed by parents toward their offspring is an exceptional one because increases in the offspring's survival directly increase the parent's current reproductive success. Any effective parental care that is costly for the parents also, simultaneously, directly increases the parents' reproductive success. However, this increment to the parents' reproductive success is not as great as the benefit to the offspring; the increment in the parents' current reproductive success must be weighed against the cost to the parents' future reproduction.

Another feature of the benefits and costs involved in parental behavior is that any cost incurred by the parent also entails a cost for the recipient or beneficiary of the act. It does so through inclusive fitness effects, or directly through reduced parental ability to continue investment or even to survive. This self-inflicted cost places limitations on the evolution of selfishness or exploitation on the part of the offspring. However, the cost to the parent, in terms of future reproduction, will be greater than the cost to the offspring, through inclusive fitness (Trivers, 1974). Because of these asymmetries in costs and benefits to each participant, a "conflict of interest" can arise between parent and offspring. This conflict can be expected to shape the family interactions that evolve.

An appreciation of these features of parent-offspring interactions is due to a considerable extent to the work of Trivers (1972, 1974), which in turn is based heavily on Hamilton's (1964) pivotal papers demonstrating the importance of a consideration of inclusive fitness. Despite the widespread references to these few papers, most basic theoretical and empirical problems await analysis, particularly as they apply to vertebrates but also for seemingly simple invertebrates (Mertz et al., 1984). As we shall see, anthropoid primates are particularly appropriate subjects, and yet often frustrating ones, for a consideration of the evolution of family interactions.

PRIMATE LIFE HISTORIES

Most anthropoid primates, meaning monkeys and apes, are relatively social animals. They live their whole lives in close proximity to other members of their species, many though not all of whom are close relatives—parents, offspring, full or half siblings. Primate groups represent a wide range of sizes, composition, and degrees of genetic relatedness.

Within, as well as among, primate species, variability in ecological conditions and a host of factors that lead to temporal and spatial variability in group composition lead to differences in the situations in which parents raise their offspring and in which these offspring continue to develop. I focus here, however, on aspects of infant care and development in which most anthropoids are similar to each other and in which they are quite different from almost all other mammals. Most monkeys and apes produce only a single young per gestation (Schultz, 1948; Leutenegger, 1979); they do so after an unusually long gestation period for animals of their size (Western, 1979); and the single offspring develops relative independence of its parents very slowly. This offspring then usually has a long juvenile period in which it still may be dependent on family members, but to a lesser extent than during infancy. All pre-reproductive stages occupy an unusually long part of the lifespan (Schultz, 1969; Eisenberg, 1981), and in most anthropoid species infants are totally dependent for survival on one or both parents for at least a year, much longer for the great apes.

Consequently, in primates there is no single brief season allotted to parental care, however intense. Rather, care of offspring is a continuing activity that must be integrated with all of an adult's maintenance activities over a long period. Moreover, most primates do not use nests or caches for their young, nor do they have communal hunting and sharing of food. A young infant and its parent, almost always its mother, are essentially "saddled" to each other, almost literally as well as figuratively, in that a very young infant rides on its mother wherever she goes. These various primate characteristics lead to the expectation that primate parents and their offspring, more than parents and young of most other species, will have evolved a complex, often subtle, finely tuned set of interactions that enable them to satisfy their highly, but not completely, overlapping interests over a prolonged period. For a mother, investment in replacement offspring is very costly if the present one dies, and for an infant, its own survival is totally dependent on the survival of its mother. Although features such as potential conflict of interest may favor evolution of behavioral conflict within families (Trivers, 1974), these other factors will favor the evolution of compromise and cooperation and have as yet received less attention (J. Altmann, 1980, 1983).

Finally, primate sibship sizes are small. Siblings are usually half siblings, particularly in multi-male groups. Maternal siblings will often differ appreciably in age whereas paternal siblings are more likely to be similar in age, members of the same age cohort (J. Altmann, 1979).

Partially because primates mature so slowly and are so long-lived, it takes a long time to obtain data that enable us even to begin an evaluation of parameters that are important to evolutionary questions. Twenty-five years after the main development of primate field studies (see survey in S. Altmann, 1967), for only a few species are we even now beginning to accumulate those necessary lifetime data. One such species, *Papio cynocephalus*, includes the savannah baboons that have been the subjects of a longitudinal research project in the Amboseli National Park of southern Kenya since 1971 (e.g., Hausfater, 1975; J. Altmann et al., 1978, 1981; Post, 1982; Walters, 1981). By focusing on the results being obtained in that study, and by then considering the ways in which these animals differ from other populations and primate species, I hope to provide insights into general features of the evolution of primate family interactions.

In the sections that follow, I first highlight those characteristics of baboons that are probably very critical to shaping the parental investment and interaction patterns to be described. Then I consider selective pressures and behaviors of potential importance at three stages of parental care and family interaction: during gestation, during infancy, and during the juvenile period. My emphasis is on intra-population variability—its nature, sources, and consequences. However, I consider, as appropriate, the interactions and investment patterns that have been observed or would be predicted under conditions that differ from those pertaining to these baboons.

BABOON LIFE HISTORIES

The methods used to obtain the Amboseli baboon data have been detailed elsewhere (see Hausfater, 1975, for dominance data; Altmann and Altmann, 1970, and J. Altmann et al., 1977, 1981, for demographic data; and J. Altmann, 1980, for family interaction data). Briefly, almost daily records are kept on all members of at least one social group within which all animals are individually identifiable through natural differences in appearance. Individuals are observed for changes in health, reproductive condition, physical maturation, and agonistic dominance relationships. These data, which are collected every year, have been complemented in some years by data from detailed studies of parental care and infant or juvenile development (e.g., J. Altmann, 1980; Stein, 1984; Pereira, 1984). In addition, mating records are available, but with varying degrees of completeness for different years. The animals are not trapped, handled, or otherwise overtly interacted with by the investigators.

Savannah baboon species live in semi-closed multi-male, multi-female groups in African grasslands and woodlands (e.g., for olive baboons, Washburn and DeVore, 1963; Rowell, 1964; Harding, 1977; Strum, 1982; for yellow baboons, Altmann and Altmann, 1970; Rasmussen, 1979; for chacma baboons, Stoltz and Saayman, 1970; Hamilton et al., 1976). The group that provides most of the Amboseli longitudinal data, named Alto's Group, has averaged about 45 animals over a 13-year period, but has had a membership as low as 30 or 35 and as high as the lower 60s. As in most primate species (Wrangham, 1980), males are the sex that disperses; with rare exceptions (reported in Rasmussen, 1981, for yellow baboons in Mikumi Park, Tanzania) baboon females remain in their natal group for life.

Baboons are highly dimorphic and, in Amboseli, females usually conceive their first infants at about six years of age, approximately two or three years before the males of their age cohort reproduce (J. Altmann et al., 1977, 1981). Mating within these multi-male, multi-female groups is semi-polygynous, semi-polyandrous; each male mates primarily with several females, each female with several males. Based on behavioral records during the few days of likely conception, infants can be assigned from one to three probable fathers, on the average two (J. Altmann et al., in press). Litter size is almost always one and most direct care of an infant is by its mother, although selective and directed male care (probably paternal care in most cases) is much greater for these multi-male primate groups than was originally assumed (see, e.g., Packer, 1979; J. Altmann, 1980; Stein, 1984).

Infant survivorship in Amboseli is slightly less than 70% for the first year (about 80% if miscarriages and stillbirths are excluded) and about 50% from birth through year two, which corresponds approximately to the interbirth interval. Survivorship is higher, and age of first reproduction lower, in situations of richer natural or human-provisioned food and of reduced predation risk (e.g., Strum and Western, 1982). If an infant dies, its mother quickly becomes pregnant again. Because mating and conceptions occur throughout the year, baboons, unlike most other primates, are not constrained to wait until a subsequent breeding season to recoup reproductive loss.

In Alto's Group, reproduction entails a mortality cost (J. Altmann, 1983). Adult females are at the highest risk of mortality while caring for young, moderate risk while pregnant, and lowest while in other reproductive stages, during menstrual cycles.

Finally, baboon and macaque species have highly stable adult female agonistic dominance hierarchies. During maturation, daughters assume their mothers' relative dominance position (Walters, 1980, and references therein; Hausfater et al., 1982). In contrast, male dominance is less affected (perhaps not at all in wild baboons) by mother's rank, and males change dominance throughout adulthood. For macaques, there is limited evidence from provisioned, semi-captive, semi-free-ranging animals (Meikle et al., 1984), but not yet any from wild ones, that maternal dominance rank affects sons' lifetime reproductive success. The body of evidence for a relationship between female dominance rank and fitness is basically, but not entirely, positive (Drickamer, 1974; Sade et al., 1977; Mori, 1979; Gouzoules et al., 1982; J. Altmann et al., in press). The weakness of the positive relationship may not be too surprising given the brevity of existing studies relative to the stability of female dominance ranks.

This, then, is the demographic and sociological backdrop for parental care and parental investment in savannah baboons. In the sections that follow, I consider various actions by parents that provide benefit to the current offspring (increase that offspring's chances of survival) and that simultaneously incur a reproductive cost (decrease the parents' ability to invest in future offspring). The difference between benefits and costs, rather than just benefit as sometimes appears in the empirical literature, is Trivers's "parental investment" (Trivers, 1972). The distinction is that parental investment is not synonymous with parental care. We are also concerned with behavior that offspring direct toward their parents, and the interaction patterns that emerge from the behavior of each.

GESTATION

Gestation is truly the first stage of both parental care and investment. Pregnancy reduces, and in some aspects terminates, care in the previous offspring and it precludes immediate investment in another one. Mammalian mothers and fetuses affect each other, behaviorally and otherwise, and our data suggest that pregnancy slightly increases a female's risk of mortality. This relatively ignored preparturitional period of investment provides some rather interesting data.

Pregnancy averages 177 days in Amboseli baboons, somewhat less time in smaller monkeys and longer in the apes (Eisenberg, 1981). During this period, the fetus of most anthropoids grows to approximately 7% of its mother's body weight (Leutenegger, 1973). Pregnant females socialize less and feed more than do cycling ones. Aside from walking enough to keep up with their group, they spend virtually all their time feeding and resting (J. Altmann, 1980; Altmann and Mututua, 1984, and unpublished data; Silk, in preparation). In Amboseli, gestation length is positively correlated with survivorship during an infant's first year. Females that are low in the dominance hierarchy tend to have slightly higher rates of spontaneous abortions, stillbirths, and short gestations (J. Altmann et al., in press). These findings raise questions about the proximal factors controlling gestation length and about the potential for female-female competition, through aggression for example, that might result in early births.

We have found that there is a sex-ratio bias by dominance rank: the highest-ranking third (ranks 1–7) of females produce three to four females to each male offspring and the lowest-ranking third produces two males to each female offspring (J. Altmann, 1980; J. Altmann et al., in press). This result has been consistent from year to year and has now been found in a second group in Amboseli as well as in several studies of captive or provisioned colony macaques (Silk et al., 1981; Simpson and Simpson, 1982), but not in all (Meikle et al., 1984). Based on a synthetic review of the literature from wild, provisioned, and captive primates, Van Schaik and Van Noordwijk (1983) conclude that social stress increases male bias in birth, but they suggest that this is not an adaptive response.

In addition to sex biases at birth, in Amboseli there is a somewhat higher survival of "right sex" offspring; that is, for high-ranking females a higher proportion of daughters (0.50) than sons (0.00) reach age six (age of first conception) and for low-ranking females, conversely, survival is higher for sons (0.33) than for daughters (0.25) (J. Altmann et al., in press). Moreover, as can be seen from these figures, daughters of high-ranking females have higher survivorship than those of low-ranking females and survivorship is higher for sons of low-ranking females than for those of high-ranking females. These differences in survivorship exaggerate the sex bias present at birth. A possible partial contribution to these survivorship differences may lie in another finding: gestation length is slightly longer for low-ranking females when they are carrying sons rather than daughters and the opposite is true for high-ranking females.

Consistent with male-biased sex ratios in low-ranking females are results of a study of captive macaques by Sackett et al. (1975) in which it was found that pregnant females received more severe aggression when they were carrying female offspring than male offspring. That finding suggests a mechanism by which low-ranking females might have shorter gestations imposed on them for female offspring, but would not contribute to an understanding of the bias against male offspring of high-ranking females. Moreover, this explanation is based entirely on the birth sex-ratio effect being one of biasing secondary sex ratios and not primary ones, whereas close examination of data for each baboon menstrual cycle allows detection of all pregnancies that lasted at least as long as implantation (7–10 days), and few miscarriages occurred in Amboseli thereafter. Consequently, we propose that at least some sex-ratio biasing occurs by the time of conception, perhaps through timing of conception with respect to ovulation and/or through factors that affect vaginal pH (e.g., Guerrero, 1974; Harlap, 1979, and subsequent commentaries; James, 1980, for evidence in humans regarding sex ratios as a function of timing within the menstrual cycle). In sum, then, Amboseli females of a given dominance rank are producing, and probably conceiving, the sex of offspring that has the best prognosis for survival and reproduction. Both the proximate and ultimate factors resulting in sex-biased investment and in particular in biased offspring sex ratios are complex. These factors are currently the topic of much attention that should result in interesting developments in the next several years.

INFANCY

By the time of parturition, a primate mother has already invested in her infant almost a third of the months that she will altogether; in baboons the six months of gestation is followed by at least twelve of care. The duration of infancy is roughly scaled to body size across monkey species and it is even longer than expected for size in apes than it is in monkeys (Schultz, 1969; Eisenberg, 1981).

In addition to the life history features already discussed, a complex of three physical characteristics of anthropoid mothers and infants seems to have a particularly strong influence on the care options during early development: a relatively "thin" low-lipid milk (Buss, 1971), locomotor altriciality, and a strong clinging/grasping response (Hines, 1942). This is not meant to imply that these features preceded the behavioral care systems in an evolutionary sense, but only that they are at present closely linked to, and developmentally may constrain, interaction possibilities. The low-lipid content of the mother's milk, combined with the fact that the amount of milk available at any one time in the mammary glands is small (Buss, 1971), means that the infant requires frequent nursing throughout most of the day during a considerable period of growth. The infant must therefore be kept relatively close to its mother most of the time. The infant's inability to orient and locomote independently necessitates that it be carried to obtain enough milk, and its well-developed grasping ability enables it to provide the active behavioral component to accomplish this proximity with minimal impediment to the mother's locomotion.

In primate species that travel little or those that live in small monogamous groupings in which individuals travel close together, the infant needn't be carried on the mother herself in order to have adequate access to her; it can be left nearby, often in the care of others in the relatively sedentary species. In most monogamous species the infant rides on its father or siblings (Kleiman and Malcolm, 1981). In most other primate species, however, access necessitates or is more efficiently effected by the infant riding on its mother, to whom it clings tenaciously. In fact, a baboon infant that gets separated from its mother during the first few weeks of life will cling tightly to whoever carries it; we have witnessed one death (Shopland, in press) and one near fatal case of infants that clung to other females who had kidnapped them, and the infants thereby hindered their mothers' ability to recover them. Before the age at which an infant recognizes its mother and selectively orients to her, the mother's ability to keep contact with the infant may be critical in some species.

Individual Differences in Style of Maternal Care

Although infants of high-ranking mothers and those of low-ranking mothers are probably exposed to fairly similar ecological dangers, it is the offspring of low-ranking females, and the mothers themselves, who are subjected to constant interest, harassment, and potential kidnapping. High-ranking mothers and their offspring are relatively immune to this interference. Apparently in response to this difference in social hazards, low-ranking mothers tend to be more restrictive and protective of their infants and to continue this restrictiveness until their infants are several months old (J. Altmann, 1978, 1980). These infants develop independence more slowly than their high-ranking peers, even after the restrictiveness is terminated.

The protective style of low-ranking mothers probably increases infant survival. Simultaneously, however, it may increase two other costs. Protectiveness and the resulting delay of independence probably results in delay of subsequent reproduction for the mother and, by delaying behavioral maturation, delays the ability of the youngster to survive independently if it should be orphaned. Our data remain suggestive on both these points, but Nicolson (1982) has demonstrated the former in a population of olive baboons. From the standpoint of natural selection, one would want to know whether restrictiveness is a stable characteristic of a female during her lifetime and whether it occurs disproportionately in her daughters. Because of the stability of dominance ranks, we predict that this is the case for baboons and are currently collecting the relevant longitudinal data. Other individual characters that affect offspring development and survival in primates await identification.

Changes in Mother-Infant Interaction Patterns During Development

Even the youngest baboon infant can maintain the position needed to obtain transport and to reach the nipple for frequent suckling, although during the first few days its mother's assistance is occasionally needed, especially toward the end of the day. All other behavior needed for the neonate's care is done by the mother, not the infant. The demand for care and attentiveness actually increases in the next month as the infant begins to break contact with its mother and explore nearby. The mother then repeatedly watches, follows, and retrieves her infant.

As the infant gets older, however, it obtains the contact and care it needs only if it contributes to the maintenance of contact and facilitates the mother's ability to provide other care, such as keeping track of its mother's location in case a predator attack occurs (J. Altmann, 1980; see also Hinde et al., 1964, and sequelae; Berman, 1980, for macaques). In addition, the infant increasingly must attend to the nature of its mother's ongoing activities—such as walking, feeding or resting—and time its contact or other care demands in such a way that these demands will be compatible rather than interfere with the mother's procurement of food (J. Altmann, 1980; see recent elegant experimental demonstrations of the effect of nutritional level on mother-infant relations in Rosenblum, 1982).

Thus, a major developmental feature of primate mother-infant interactions is the increasing role that the infant must play in facilitating its own care. It must become sensitive to its mother's activities and make responses that are contingent on her activities. Moreover, these contingencies change during development with alterations in the infant's size, needs, potential hazards, and so forth. Consequently, for a primate infant a premium is placed on development of social sensitivity, changing contingent responses, and coordinated, collaborative activity with a social partner. Developmentally, these interactions probably form the basis for the complex, flexible primate behavior as we know it, the most intense evolutionary pressure for which may have come as a result of severe ecological pressures on primate mothers (J. Altmann, 1980, 1983).

Independent Manipulation of Benefit and Cost in Maternal Care

One of the consequences of temporal overlap between parental activities and maintenance activities such as feeding is that a given level of benefit to an infant, say that provided by a milliliter of milk, might be obtained at quite different costs to the mother, depending on her other activity at the time. If the suckling occurs when she is foraging, the suckling of a large infant probably reduces her feeding efficiency. In contrast, the suckling does not entail this extra cost if it occurs when the mother is resting. This possiblility of providing the same benefit at lower cost probably is one of the factors leading to the contingent responses described above.

There is an additional way, suggested by the work of Konner and Worthman (1980) with humans, that benefits might be delivered at two different costs. A human mother who suckles her infant, even briefly, at least once every hour will maintain high prolactin levels and experience longer post-partum amenorrhea than one who provides the same amount of suckling time and milk but distributed such that a number of long periods occur without suckling. It remains to be determined if other lactating primates, including baboons, respond in the same way to inter-nursing intervals. Of course, this example is not as simple as it at first appears because the two infants will obtain the same benefit in the short run whereas one will probably experience termination of nursing at a younger age. Nonetheless, the example serves as a striking reminder that we must distinguish costs and benefits, and be mindful of that separation and of its implications for the evolution of complex family interaction patterns.

Paternal Care

In most primate species other than the monogamous, primarily twin-bearing ones, care is overwhelmingly by the infant's mother, partially for reasons suggested above. However, even in multi-male social systems such as those found in baboons, some care is provided by adult males and each male directs care primarily toward infants that are more likely to be his offspring. This selectivity has now been reported from several studies of anubis, chacma, and cynocephalus baboons (J. Altmann, 1978, 1980; Packer, 1979; Smuts, 1982; Busse, 1984; Stein, 1984).

Although genetic information was not available in any of the afore-mentioned studies, the baboons' consortship mating system and the external indicators of ovulation-timing (Gillman and Gilbert, 1946; Kriewaldt and Hendrickx, 1968) make it likely that the indicators of paternity that are available to human observers (as well as to the baboons) are fairly reliable if they are recorded in detailed, systematic records on a daily basis (J. Altmann et al., in press). It seems that persistent adult male-female bonds and consortships (e.g. Hausfater, 1975; Rasmussen, 1980; Smuts, 1982) provide much of the proximal cues for male care of those specific infants that are likely to be their own. The only series of experimental studies of anthropoid ability to identify paternal kin in the absence of experiential cues produced somewhat ambiguous results in macaques (Wu et al., 1980; Fredrickson and Sackett, 1984).

The amount of male parental care is variable at every level of organization. Variability among primate species is great even within a single genus. Within *Papio* (baboons), for example, those species with multi-male groups seem to exhibit much more direct male care than do hamadryas baboons which live primarily in single-male groups. Within *Macaca* there is considerable variability in paternal care from species to species, even among species with relatively similar social and mating systems (see Snowdon and Suomi, 1982, for a recent cross-species review). Although there still is a relative paucity of quantitative data on primate paternal care (but see chapters in Taub, 1984), it is illuminating to consider the kinds of care that males provide infants and how this care changes during development.

In monogamous, primarily twin-bearing small anthropoids, fathers and older siblings provide virtually all non-nutritional care for infants. This care probably is critical in allowing higher rates of reproduction in these species than would otherwise be expected for primates of their size (Kleiman, 1977; Ralls, 1977). In baboons, the non-monogamous species in which male care has been best documented, the care is of several sorts, usually, but not always, complementary to maternal care rather than a substitute for it (Snowdon and Suomi, 1982), especially in the case of young infants.

One of the first roles that a baboon male plays is that of social buffer for the mother-infant dyad. In particular the male reduces the harassment of mothers that are low on the dominance hierarchy by those that are high-ranking (J. Altmann, 1980). This social buffering reduces the stress on mothers, allows their infants to explore more, and seems to facilitate the mothers' ability to feed undisturbed. Some mothers even forage away, leaving their infants with these male "babysitters" or "care-providing fathers," as the case may be. These males, unlike adult females and juveniles, rarely try to carry a young infant in this situation; rather, they sit and watch it closely, give repeated soft grunts, and follow the infant if it moves.

As the young baboon infant begins to feed on solid food, it gains a second advantage from its proximity to the male. The infant, unlike other group members, is tolerated immediately next to the male as he feeds. The infant thereby has the opportunity to learn about food sources from a second adult and also has access to scraps of food that would be costly or impossible for the infant to obtain on its own. These are usually plant food sources, but scraps of vertebrate meat are also sometimes obtained this way by those infants whose male caregivers are frequent predators.

One of the first areas in which a baboon mother rejects her infant's attempts to obtain care is in providing transport during the group's foraging and progressions. Infants often show clear signs of fatigue by mid-afternoon, particularly if the day journey has been long or the afternoon is quite hot. At such times the infant's associated adult male will often allow the infant to ride when the mother refuses to do so. At this stage, a mother will still carry her infant in the case of some external danger such as a predator. Soon, however, the growing infant's weight sufficiently slows its mother that carrying probably becomes disadvantageous to both, even though the infant cannot keep up with the group in flight. It is at these times when the male's role is particularly visible and dramatic as he dashes back and retrieves a stranded youngster.

As the older infant gradually becomes a young juvenile and its mother begins to invest in her next offspring, another important male behavior emerges. Young juveniles begin to be the object of aggression by older animals or by those peers whose mothers are dominant to their mothers. The adult male, like the mother, will sometimes come to the aid of the youngster, primarily if the harassment escalates or persists. Because the male is dominant to all adult females and immature animals, he can provide effective support against more antagonists than can the mother.

JUVENILE PERIOD

No sharp transition separates infancy from the juvenile period. The age at which the juvenile stage has begun is that at which a youngster is nutritionally independent, the youngster could survive its mother's death, and its mother has probably begun investment in her next offspring (Pereira and Altmann, 1985).

A surviving monkey mother, however, still provides some care, an appreciable amount in the case of the great apes (see, e.g., Pusey, 1983). In Amboseli, the young juvenile baboon usually still sleeps nestled in its mother's lap in the trees at night until the birth of the next infant. Thereafter, the juvenile sleeps against its mother's body outside her ventrum or huddles with its adult male associate or an older sibling. A juvenile sometimes continues to receive agonistic support from its likely father, if he is still present in the group. A mother also continues to provide some agonistic support for her juvenile offspring (Walters, 1980, and references therein for various macaque and baboon species; Pereira, 1984), but low-ranking mothers are less able and perhaps less willing to do so. As juveniles mature, they not only receive agonistic aid from their mothers but they in turn come to the aid of their mothers, forming agonistic coalitions with them. Even among animals of similar ranks, individual and family differences seem to occur in the amount of agonistic support, but these differences are as yet undocumented quantitatively. Family size seems to be a relevant factor, as would be expected for the female kin-based groups such as found in baboons and macaques (Silk and Boyd, 1983).

Perhaps the major area of care and of social interaction between a mother and her juvenile offspring involves social grooming. In his study of social development in baboon juveniles, Pereira (1984; Pereira and Altmann, 1985) found that a young juvenile receives over 50 times more grooming from its mother than, per capita, from other females. Moreover, a mother grooms her young juvenile son or daughter approximately nine times as much as the juvenile grooms her. Thus, the benefits of grooming are still provided for a young juvenile predominantly by its mother and grooming interactions between them are still characterized by an asymmetric, care-giving quality. This relationship changes in interesting ways during maturation. By the time a juvenile is three years old, it appears that the mother expects not to be providing care, but to have a reciprocally altruistic relationship with her offspring. Male offspring, most of whom will eventually disperse from their natal group and spend their adult years in other social groups, continue to provide little grooming. In addition, mothers now refuse sons' grooming solicitations and do not initiate grooming of their sons. Consequently, few grooming interactions occur between mothers and older juvenile sons. Although the grooming interactions of older juvenile daughters are less exclusively with their mothers than was the case earlier, mothers are still their daughters' predominant adult female grooming partners. In addition, the grooming between mother and daughter is symmetric, each sharing equally the role of actor and recipient.

Thus, during the juvenile period we see a gradual shift in parent-offspring interactions from an asymmetric, parental care-giving pattern to one characterized by reciprocal altruism reinforced by kinship, a pattern that mothers and daughters will retain during adulthood in their natal groups. Sons are not likely to be future partners in reciprocal relationships. For them, the patterns of parent-offspring interactions wane. The skills and sensitivities that have been developed will have other applications for these youngsters who as adults will need to develop all their social relationships anew in another group.

CONCLUSIONS

Primate family interactions are intense and extend over a long period. Especially for mothers and daughters and also for sisters of most species, these close relationships extend beyond the developmental stages discussed above, throughout adulthood. The complexities of these relationships have only recently been investigated quantitatively, and then for only a very few species and topics. Paternity information is less accessible than is maternity determination, and it is males that disperse in most species, including the best-studied species. Consequently, we are particularly ignorant of the life course of relationships for males, as we are for most aspects of male life histories, even for the best-studied species.

In addition, the existing data dramatically demonstrate the importance of quantitative studies that continue to delve below the surface, that are longitudinal rather than just cross-sectional, and that include investigation of variability within species and within populations as well as that between species. Serious attention to primate flexibility and within-lifetime adaptability to diverse conditions must be incorporated in investigations of this taxonomic group. Finally, we can expect recent economic and population models of the evolution of family interactions to continue to enrich the study of primate behavior, but the utility of the models, and their ultimate growth, will also depend on their ability to incorporate features of the life histories of long-lived, slowly producing and developing species such as primates.

ACKNOWLEDGMENTS

My research in Kenya is conducted under the sponsorship of the Institute of Primate Research, National Museums of Kenya. Thanks are due to the Institute and to the other agencies and individuals in Kenya who facilitate our work there. My colleagues, Stuart Altmann and Glenn Hausfater, have shared the directorship of the project and I am grateful to them and to all the Amboseli baboon researchers who have contributed to the basic monitoring data during the past 13 years. The Amboseli Baboon Research Project has benefited from the financial support of the Harry Frank Guggenheim Foundation, the Spencer Committee of the University of Chicago, the National Institutes of Health, and the National Science Foundation. Carolyn Johnson assisted with manuscript preparation. Stuart Altmann, Lynne Houck, and three anonymous reviewers provided helpful comments on an earlier version of the manuscript.

LITERATURE CITED

Altmann, J. 1978. Infant independence in yellow baboons. In: Burghardt, G., and M. Bekoff (eds), The Development of Behavior. Garland STPM Press, New York, pp. 253–277.

Altmann, J. 1979. Age cohorts as paternal sibships. *Behavioral Ecology and Sociobiology* 6:161–164.

Altmann, J. 1980. Baboon Mothers and Infants. Harvard University Press, Cambridge, Mass.

Altmann, J. 1983. Costs of reproduction in baboons. In: Aspey, W. P., and S. I. Lustick (eds), Behavioral Energetics: Vertebrate Costs of Survival. Ohio State University Press, Columbus, pp. 67–88.

Altmann, J., S. Altmann, and G. Hausfater. 1978. Primate infant's effects on mother's future reproduction. *Science* 201:1028–1029.

Altmann, J., S. Altmann, and G. Hausfater. 1981. Physical maturation and age estimates of yellow baboons, *Papio cynocephalus. American Journal of Primatology* 1:389–399.

Altmann, J., S. Altmann, and G. Hausfater. In press. Determinants of reproductive success in savannah baboons *(Papio cynocephalus)*. In: Clutton-Brock, T. (ed), Reproductive Success. University of Chicago Press, Chicago.

Altmann, J., S. Altmann, G. Hausfater, and S. A. McCuskey. 1977. Life history of yellow baboons: Physical development, reproductive parameters, and infant mortality. *Primates* 18:315–330.

Altmann, J., and R. S. Mututua. 1984. Determinants of activity budgets and spatial affinities of adult female baboons *(Papio cynocephalus)* in Amboseli National Park, Kenya. *International Journal of Primatology* 5:256 (abstract).

Altmann, S. 1967. Preface. In: Altmann, S. A. (ed), Social Communication Among Primates. University of Chicago Press, Chicago, pp. ix–xii.

Altmann, S. A., and J. Altmann. 1970. Baboon ecology: African field research. S. Karger, Basel, and University of Chicago Press, Chicago.

Berman, C. M. 1980. Mother-infant relationships among free-ranging rhesus monkeys on Cayo Santiago: A comparison with captive pairs. *Animal Behaviour* 28:860–873.

Buss, D. H. 1971. Mammary glands and lactation. In: Hafez, E. S. E. (ed), Comparative Reproduction of Nonhuman Primates. Thomas Press, Springfield, Ill., pp. 315–333.

Busse, C. 1984. Triadic interactions among male and infant chacma baboons. In: Taub, E. M. (ed), Primate Paternalism. Van Nostrand Reinhold, New York, pp. 186–212.

Drickamer, L. C. 1974. A ten-year summary of reproductive data for free-ranging *Macaca mulatta. Folia Primatologica* 21:61–80.

Eisenberg, J. F. 1981. The Mammalian Radiations. University of Chicago Press, Chicago.

Fredrickson, W. T., and G. P. Sackett. 1984. Kin preferences in primates *(Macaca nemestrina)*: Relatedness or familiarity? *Journal of Comparative Psychology* 98:29–34.

Gillman, J., and C. Gilbert. 1946. The reproductive cycle of the chacma baboon *(Papio ursinus)* with special reference to the problems of menstrual irregularities as assessed by the behavior of the sex skin. *South Africa Journal of Medical Science* 11 (Biological Supplement):1–54.

Gouzoules, H., S. Gouzoules, and L. Fedigan. 1982. Behavioural dominance and reproductive success in female Japanese monkeys *(Macaca fuscata). Animal Behaviour* 30:1138–1150.

Guerrero, R. 1974. Association of the type and time of insemination within the menstrual cycle with the human sex ratio at birth. *New England Journal of Medicine* 291:1056–1059.

Hamilton, W. D. 1964. The genetical evolution of social behavior, I, II. *Journal of Theoretical Biology* 7:1–52.

Hamilton, W. J. III, R. E. Buskirk, and W. H. Buskirk. 1976. Defense of space and resources by chacma *(Papio ursinus)* baboons in an African desert and swamp. *Ecology* 57:1264–1272.

Harding, R. S. O. 1977. Patterns of movement in open country baboons. *American Journal of Physical Anthropology* 47:349–353.

Harlap, S. 1979. Gender of infants conceived on different days of the menstrual cycle. *New England Journal of Medicine* 300:1445–1448.

Hausfater, G. 1975. Dominance and reproduction in baboons: A quantitative analysis. *Contributions to Primatology* 7. S. Karger, Basel.

Hausfater, G., J. Altmann, and S. Altmann. 1982. Long-term consistency of dominance relations among female baboons *(Papio cynocephalus). Science* 217:752–755.

Hinde, R. A., T. E. Rowell, and Y. Spencer-Booth. 1964. Behaviour of socially living Rhesus monkeys in their first six months. *Journal of Zoology* 143:609–649.

Hines, M. 1942. The development and regression of reflexes, postures and progression in the young macaque. *Contributions to Embryology* 196:155–209.

James, W. H. 1980. Time of fertilisation and sex of infants. *Lancet* 1:1124–1126.

Kleiman, D. G. 1977. Monogamy in mammals. *Quarterly Review of Biology* 52:39–69.

Kleiman, D. G., and J. R. Malcolm. 1981. Evolution of male parental investment in mammals. In: Gubernick, D. J., and P. H. Klopfer (eds), Parental Care in Mammals. Plenum Press, New York, pp. 347–387.

Konner, M., and C. Worthman. 1980. Nursing frequency, gonadal function, and birth spacing among !Kung hunter-gatherers. *Science* 207:788–791.

Kriewaldt, F. H., and A. G. Hendrickx. 1968. Reproductive parameters of the baboon. *Laboratory Animal Care* 18:361–370.

Leutenegger, W. 1973. Maternal-fetal weight relationships in primates. *Folia Primatologica* 20:280–293.

Leutenegger, W. 1979. Evolution of litter size in primates. *American Naturalist* 114:525–531.

Meikle, D. B., B. L. Tilford, and S. H. Vessey. 1984. Dominance rank, secondary sex ratio, and reproduction of offspring in polygynous primates. *American Naturalist* 124:173–188.

Mertz, D., D. M. Craig, M. J. Wade, and J. F. Boyer. 1984. Cohort selection. *Evolution* 38:560–570.

Mori, A. 1979. Analysis of population changes by measurement of body weight in the Koshima troop of Japanese monkeys. *Primates* 20:371–397.

Nicolson, N. 1982. Weaning and the development of independence in olive baboons. Ph.D. thesis. Harvard University, Cambridge, Mass.

Packer, C. 1979. Male dominance and reproductive activity in *Papio anubis*. *Animal Behaviour* 27:37–46.

Pereira, M. E. 1984. Age changes and sex differences in the social behavior of juvenile yellow baboons *(Papio cynocephalus)*. Ph.D. thesis, University of Chicago, Chicago.

Pereira, M. E., and J. Altmann. 1985. Development of social behavior in free-living nonhuman primates. In: Watts, E. S. (ed), Nonhuman Primate Models for Growth and Development. Alan R. Liss, New York, pp. 217–309.

Post, D. G. 1982. Feeding behavior of yellow baboons *(Papio cynocephalus)* in the Amboseli National Park, Kenya. *International Journal of Primatology* 3:403–430.

Pusey, A. E. 1983. Mother-offspring relationships in chimpanzees after weaning. *Animal Behaviour* 31:363–377.

Ralls, K. 1977. Sexual dimorphism in mammals: Avian models and unanswered questions. *American Naturalist* 111:917–937.

Rasmussen, D. R. 1979. Correlates of patterns of range use of a troop of yellow baboons *(Papio cynocephalus)*. I. Sleeping sites, impregnable females, births and male emigrations and immigrations. *Animal Behaviour* 27:1098–1112.

Rasmussen, D. R. 1981. Communities of baboon troops (*Papio cynocephalus*) in Mikumi National Park, Tanzania: A preliminary report. *Folia Primotologica* 36:232–242.

Rasmussen, K. L. 1980. Consort behavior and mate selection in yellow baboons *(Papio cynocephalus)*. Ph. D. thesis, Cambridge University, Cambridge.

Rosenblum, L. A. 1982. The influence of the social and physical environment on mother-infant relations. *Annali del Instituto Superiore di Sanita* 18(2):215–222.

Rowell, T. E. 1964. The habit of baboons in Uganda. *Proceedings of the East African Academy* 2:121–127.

Sackett, G., R. Holm, and S. Landesman-Droyer. 1975. Vulnerability for abnormal development: Pregnancy outcomes and sex differences in macaque monkeys. In: Ellis, N. R. (ed), Aberrant Development in Infancy, Human and Animals Studies. John Wiley, New York, pp. 59–76.

Sade, D. S., K. Cushing, P. Cushing, P. Dunaif, A. Figueroa, J. Kaplan, C. Lauer, D. Rhodes, and J. Schneider. 1977. Population dynamics in relation to social structure on Cayo Santiago. *Yearbook of Physical Anthropology* 20:253–262.

Schultz, A. H. 1948. The number of young at birth and the number of nipples in primates. *American Journal of Physical Anthropology* 6 N.S.:1–23.

Schultz, A. H. 1969. The Life of Primates. Universe Books, New York.

Silk, J. B., and R. Boyd. 1983. Cooperation, competition, and mate choice in matrilineal macaque groups. In: Wasser, S. K. (ed), Female Vertebrates. Academic Press, New York, pp. 316–349.

Silk, J. B., C. B. Clark-Wheatley, P. S. Rodman, and A. Samuels. 1981. Differential reproductive success and facultative adjustment of sex ratios among captive female bonnet macaques *(Macaca radiata)*. *Animal Behaviour* 29:1106–1120.

Simpson, M. J. A., and A. E. Simpson. 1982. Birth sex ratios and social rank in rhesus monkey mothers. *Nature* 300:440–441.

Smuts, B. 1982. Special relationships between adult male and female olive baboons *(Papio anubis)*. Ph.D. thesis. Stanford University, Stanford, Calif.

Snowdon, C. T., and S. J. Suomi. 1982. Paternal behavior in primates. In: Fitzgerald, H. E., J. A. Mullins, and P. Gage (eds), Primate Behavior and Child Nurturance, Vol. 3. Plenum Press, New York, pp. 63–108.

Stein, D. M. 1984. Ontogeny of infant-adult male relationships during the first year of life for yellow baboons *(Papio cynocephalus)*. In: Taub, D. M. (ed), Primate Paternalism. Van Nostrand Reinhold, New York, pp. 213–243.

Stoltz, L. P., and G. S. Saayman. 1970. Ecology and behaviour of baboons in the Northern Transvaal. *Annals of the Transvaal Museum* 26:99–143.

Strum, S. C. 1982. Agonistic dominance in male baboons: An alternative view. *International Journal of Primatology* 3:175–202.

Strum, S. C., and J. Western. 1982. Variations in fecundity with age and environment in olive baboons *(Papio anubis)*. *American Journal of Primatology* 3:61–76.

Taub, D. M. (ed). 1984. Primate Paternalism. Van Nostrand Reinhold, New York.

Trivers, R. L. 1972. Parental investment and sexual selection. In: Campbell, B. (ed), Sexual Selection and the Descent of Man. Aldine-Atherton, Chicago, pp. 136–179.

Trivers, R. L. 1974. Parent-offspring conflict. *American Zoologist* 14:249–264.

Van Schaik, C. P., and M. A. Van Noordwijk. 1983. Social stress and the sex ratio of neonates and infants among non-human primates. *Netherlands Journal of Zoology* 33:249–265.

Walters, J. 1980. Interventions and the development of dominance relationships in female baboons. *Folia Primatologica* 34:61–89.

Walters, J. 1981. Inferring kinship from behaviour: Maternity determinations in yellow baboons. *Animal Behaviour* 29:126–136.

Washburn, S. L., and I. DeVore. 1963. Baboon ecology and human evolution. In: Howell, F. C., and F. Bourliere (eds), African Ecology and Human Evolution. Aldine, Chicago, pp. 335–367.

Western, D. 1979. Size, life history and ecology in mammals. *African Journal of Ecology* 20:185–204.

Wrangham, R. W. 1980. An ecological model of female-bonded primate groups. *Behaviour* 75:262–300.

Wu, H. M. H., W. G. Holmes, S. R. Medina, and G. P. Sackett. 1980. Kin preference in infant *Macaca nemestrina*. *Nature* 285:225–227.

Index